# THE
# ENVIRONMENTAL
# ADDRESS BOOK

# Other books by Michael Levine

*The Address Book:*
*How to Reach Anyone Who* Is *Anyone*

*The Corporate Address Book*

*The Music Address Book*

# THE ENVIRONMENTAL ADDRESS BOOK

•

How to Reach the Environment's
Greatest Champions and
Worst Offenders

•

# Michael Levine

Foreword by Ed Begley, Jr.,
and Olivia Newton-John

A PERIGEE BOOK

Perigee Books
are published by
The Putnam Publishing Group
200 Madison Avenue
New York, NY 10016

Library of Congress Cataloging-in-Publication Data

Levine, Michael, date.
    The environmental address book: how to reach the environment's
greatest champions and worst offenders / Michael Levine: foreword
by Ed Begley, Jr., and Olivia Newton-John.
        p.      cm.
    Summary: Contains over 2,000 names and addresses of organizations,
agencies, celebrities, political figures, and businesses (local,
state, national, and international level) concerned with the state
of the world's environment.
    ISBN 0-399-51660-3 (pbk.: alk. paper)
    1. Environmental protection—Information services—Directories.
2. Environmentalists—Directories.   [1. Environmental protection—
Information services—Directories.   2. Environmentalists—
Directories.]   I. Title
TD169.6.L48      1991              91-10985 CIP AC
363.7'0025—dc20

Cover design © 1991 by Richard Rossiter

Printed in the United States of America
1 2 3 4 5 6 7 8 9 10

This book is printed on acid-free and recycled paper.
∞

# ACKNOWLEDGMENTS

*Few authors throughout human history leave a lasting imprint on the world, even in a small way, and yet it is almost a universal goal of any writer.*

*Rachel Carson did.*

*In 1962, she wrote a powerfully grim book about the environment called* Silent Spring *that became a best-seller and interested a young American named John F. Kennedy, who happened to be president of the United States.*

*The president urged federal agencies to respond to the book's warning of the emerging environmental crisis.*

*Ironically, at the same time Ms. Carson died of cancer at the age of fifty-six, the U.S. Senate first heard testimony about fish being killed by insecticides. A then-new environmental agency banned DDT eight years later and the world has not quite been the same since, thanks in large part to Rachel Carson.*

I'm lucky. I get to say publicly to special people in my life, many whom I love, how much they mean to me.

My literary agent and friend, Alice Martell, and her assistant, Paul Raushenbush.

My friends at Putnam, Laura Shepherd and Eugene Brissie.

My father, Arthur O. Levine, and Marilyn Beck.

My special friends, Bart Andrews, Rana Bendixen, Ken Bostic, Bill Calkins, Susan Gauthier, Karen L'Heureux, Richard Imprescia, Bette Geller Jackson, Richard Lawson, Nancy Mager, John McKillop, Julie Nathanson, Lynn Novatt, Dennis Prager, Bonnie Reiss, Owen Rutledge, Joshua Trabulus, and Erline White.

My business partners, Mitchell Schneider and Monique Moss.

My office family, Gabrielle Abrams, Jeff Albright, Amanda Cagen, Marla Capra, Kathie Collins, Kim Kaiman, Karen Lindstrom, Debbie Mindle, Suzette Mir, Stace Nelson, Tresa Redburn, Marcee Rondan, Jane Singer, Jeff Sullivan, Laura Westafer, Julie Wheeler, and Staci Wolfe.

My business associates, Laura Herlovich, Barry Langberg, Pierre Lehu, Dan Pine, Myrna Post, Diana Reisdorf, and Joy Sapieka.

And to M. Scott Peck for teaching me that "life is difficult."

Special thanks to Sal Manna and Kathleen Connor for their incredible commitment to excellence in the researching of this book.

For Shawn . . .

# CONTENTS

# FOREWORD

Congratulations! You're probably already recycling your cans, newspapers, and glass containers. By now you've probably stopped buying products harmful to the environment, such as ozone-depleting Styrofoams and wasteful plastic packaging, and maybe you're walking or cycling more and driving your car less.

Good. Keep it up. Now it's time for some broader action. You've heard us and other environmentalists say for years that saving the planet begins at home. That's still true. But it doesn't end there. Once you've integrated into your life a greener, more environmentally conscious lifestyle, you must then takes steps to see that large-scale changes are implemented. That's what this book is all about.

Michael Levine has put together in these pages a manual for action. This handbook is about power, and it is meant to be *used*. Never underestimate the impact of a single letter to influence corporate decisions, sway public officials, or frighten environmental polluters. You have at your fingertips all you need to make the most significant difference you can. By speaking up, you will be heard.

If you've been in tune with the environmental movement over the past few years, then you know time is running out. We face a terrifying array of potential calamities: greenhouse warming, rain forest destruction, ozone depletion, air and water pollution, rapidly disappearing species, toxic waste—it's almost too much for one person to comprehend.

But comprehend we must—and soon—if we're going to save this planet for ourselves, our children, and all life on Earth. Clearly, politicians and captains of industry have been woefully slow to wake up to the crisis. It's up to people like you to seize the initiative.

So write to the United Nations Environment Programme. Write to the Sierra Club. Write to the president of that multinational corporation torching the Brazilian rainforest and tell him how you feel. Keep track of your letters and responses. Urge your friends to do the same. We believe personal direct action is the key to turning the corner on the global environmental crisis. It's up to you.

Lastly, we admire your courage. It's so easy for people to look the other way and leave the problem for someone else to handle. By purchasing this book, you've made a profound statement. You are saying you will and you must make a difference in this world. That's not easy to do. But thank God you're willing to try.

Remember, don't give up. It may be discouraging at times when the polluters continually resist change, when politicians seem overly indebted to corporate fat cats, and when the sheer magnitude of the global crisis seems overwhelming. But we must pull together and persevere. In the end, we will win . . . or, more accurately, Mother Earth will win.

Ed Begley, Jr.

Olivia Newton-John

# AUTHOR'S NOTE

"... in a democracy every citizen, regardless of his interest in politics, holds office. Every one of us is in a position of responsibility, and in the final analysis, the kind of Government we get depends upon how we fulfill those responsiblities. We, the people, are the boss, and we will get the kind of political leadership, be it good or bad, that we demand and deserve. These problems do not even concern politics alone for the same basic choice of courage or compliance continually faces us all, whether we feel the anger of constituents, friends, a Board of Directors, or our union."

John F. Kennedy
Profiles in Courage

When it concerns our environment, I firmly believe that the decisions you make in your own house are more important than those made in the White House. The condition of the environment demands more than cocktail party chatter—it's time to take action. Hence, I call this work "an action book."

Active citizenship is fundamental to American life. As Alexis de Tocqueville explained over 150 years ago in *Democracy in America*, the genius of America lies not in its government but in its free associations— churches and synagogues, neighborhoods, and voluntary local activities.

The extensive research I have done for this book has taught me that the opportunities to make genuine, noticeable improvements on all areas of our environment exist generally in our own personal realm. Of course, we will need the assistance of business and government but that, too, lies within each active person's reach.

My dream is that each person who purchases this book will use it to lobby, inform, congratulate, learn from, and interact with as many people as possible to make a positive impact on our environment. In the process, you will be making an equally positive impact on yourself.

Michael Levine
Los Angeles,
California

# BAD GUYS/GOOD GUYS

This chapter includes the addresses of organizations, companies, and individuals noteworthy for being environmental offenders—the Bad Guys. It also lists some of the heroes—the Good Guys—who are helping to save the environment through a variety of media, and finally, a combination of the two—organizations, companies, and individuals whose actions have had both positive and ill effects—the Bad Guys/Good Guys. It is a hopeful sign indeed that the list of Good Guys is far longer that that of Bad Guys, a situation that would probably have been reversed only a few years ago. Hopefully the list of Bad Guys/Good Guys is an indication of how swiftly people are responding to pressure and becoming committed to helping the environment.

## BAD GUYS

**American Fur Industry**
262 7th Avenue—7th Floor
New York, NY 10001
(212) 564-5133
*Sandra Blye, Executive Vice-President*

Promotes use of animal fur for decorative wearing apparel.

**Anderson, Warren M.**
Old Ridgebury Road
Danbury, CT 06817
(203) 794-2000

CEO of Union Carbide Corporation who has overseen the company during and after the Bhopal tragedy.

**Champion International**
One Champion Place
Stamford, CT 06921
(203) 358-7000
*Andrew C. Sigler, President*

Paper company whose Canton, North Carolina, plant pumps dioxin-tainted effluent into the now-polluted Pigeon River.

**Earth First!**
P.O. Box 5871
Tucson, AZ 85703
(602) 622-1371
*Mike Roselle, Founder*

Militant environmental activist group whose actions have represented the radical end of the pro-earth movement, occasionally discrediting more appropriate activities.

**Hurwitz, Charles**
Pacific Lumber Company
10880 Wilshire Boulevard, Suite 1600
Los Angeles, CA 90024
(213) 474-6264

Lumber company CEO accused of destroying stands of redwoods for short-term cash.

**Oregon Lands Coalition**
280 Court, NE
Salem, OR 97204
(503) 363-8582
*Jackie Lang, Spokesperson*

Pro-logger group that lobbied against the protection of the spotted owl and has protested the pro-environment donations of companies such as Mattel Toys.

### Rawl, Lawrence G., CEO

Exxon Corporation
1251 Avenue of the Americas
New York, NY 10020
(212) 333-6900

Responsible for the Exxon *Valdez* oil spill and the subsequent mishandling of both the public relations and the actual cleanup.

### Rockwell International

2230 East Imperial Highway
El Segundo, CA 90245
(213) 647-5000
*Donald R. Beall, CEO*

Responsible for pollution at the Stringfellow Acid Pits in California, nuclear weapons plants in Hanford, Washington, Rocky Flats, Colorado, and elsewhere.

### Waste Management, Inc.

3003 Butterfield Road
Oak Brook, IL 60521
(708) 572-8800

Operates some of the country's biggest and worst-offending garbage and hazardous waste dumps. At one time, Waste Management, Inc., was paying fines approaching $1 million a month for landfill violations.

# BAD GUYS/GOOD GUYS

### Bumble Bee Seafoods

5775 Roscoe Court
San Diego, CA 92123
(619) 560-0404
*Mark A. Koob, President*

Bad Guys for engaging in tuna fishing that harms dolphins but now Good Guys for committing to dolphin-safe practices.

### Federal Cartridge Company

900 Ehlen Drive
Anoka, MN 55303
(612) 422-2577
*William Stevens, Manager,
Conservation Activities*

Believe it or not, an ammunition manufacturer with a conservation department.

### Hope, Bob

c/o Joe Goldstein
505 Eighth Avenue
New York, NY 10018

Bad Guy for first trying to sell off thousands of acres in the Santa Monica Mountains to developers. Good Guy for now swapping most of the land to go to a conservancy.

### Keller, George M., Chairman

Chevron Corporation
225 Bush Street
San Francisco, CA 94104
(415) 894-7700

Good Guys for protecting the land and wildlife around their drilling sites and pipe lines. Bad Guys for shamelessly promoting themselves for doing it.

### McDonald's Corporation

One McDonald Plaza
Oak Brook, IL 60521
(708) 575-3000
*Fred L. Turner, CEO*

Good Guys for promising to phase out use of Styrofoam "clam shell" food packaging. Bad Guys for taking so long.

### Pons, B. Stanley

University of Utah
Salt Lake City, UT 84112

Cold fusion scientist. Good Guy for research which promised unlimited clean energy. Bad Guy for making claims which weren't scientifically substantiated.

## Simpson, Homer

c/o Fox Television
10201 W. Pico Boulevard
Los Angeles, CA 90035

Bad Guy for not doing such a great job at the Springfield nuclear facility. Good Guy for reminding us of our human failings.

# GOOD GUYS

## Agran, Larry

P.O. Box 19575
Irvine, CA 92713
(714) 724-6205

Mayor of Irvine, California. He initiated the toughest anti-toxic local laws in the country in 1989.

## Ben and Jerry's Homemade Ice Cream

P.O. Box 240
Waterbury, VT 05676
(802) 244-5641
*Ben Cohn and Jerry Greenfield, Founders*

Environmentally aware entrepreneurs with special emphasis on raising funds to save the rain forests.

## Boycott McDonald's Coalition

P.O. Box 1836
Boston, MA 02205
*Heather Schofield, President*

Initiated the boycott which helped force McDonald's to eliminate CFC packaging.

## Brand, Stewart

27B Gate 5 Road
Sausalito, CA 94965

Author, editor, and publisher of *The Whole Earth Catalog.*

## Brower, David Ross

Earth Island Institute
Fort Mason Center
San Francisco, CA 94123
(415) 788-3666

Environmentalist who led the Sierra Club in the fifties and sixties before founding Friends of the Earth. He currently heads Earth Island Institute.

## Center for Law in the Public Interest

11835 West Olympic Boulevard, Suite 1155
Los Angeles, CA 90064
(213) 470-3000
*Ruthann Lehrer, Executive Director*

Represents groups without charge in matters of general public importance. Litigates class action cases in environmental and land use areas.

## Citizen's Call

P.O. Box 1722
Cedar City, UT 84720
(801) 268-0186

Helps radiation victims, particularly those affected by the Nevada test site, and has started a hospice for them.

## Citizen's Energy Corporation

530 Atlantic Avenue
Boston, MA 02210
(617) 951-0400
*Michael Kennedy, Director*

Includes the Walden Pond Project which is trying to raise money to purchase Walden Woods.

## Coalition for Environmentally Responsible Economics

711 Atlantic Avenue
Boston, MA 02111
(617) 451-0927

Businesses which are environmentally responsible.

## Coca-Cola Co.

One Coca-Cola Plaza, NW
Atlanta, GA 30313
(404) 676-2121
*Donald R. Keough, President*

The first soft drink company to use recycled plastic in its bottles.

## Commoner, Barry
Queens College Center for Biology
and Natural Systems
Flushing, NY 11367
(718) 997-5000

Influential biologist and environmentalist, author of *Making Peace with the Planet.*

## Consumer Reports
P.O. Box 2886
Boulder, CO 80322
(800) 234-1645
*Irwin Landav/Eileen Denver, Editors*

Publishes magazine rating the quality of consumer products.

## Eddie Bauer, Inc.
15010 NE 36th
Redmond, WA 98052
(206) 882-6100

Outdoor clothing manufacturer who has instituted the Heroes of the Earth award.

## Flooks, Ian
321 Fulham Road
London SW10 9QL England
1 352-8140
1 352-2762 FAX

Rock 'n' roll artist manager who initiated *Rainbow Warriors,* the 1989 record album to benefit Greenpeace.

## Gale, Robert Peter, Dr.
c/o UCLA Medical Center, Center
for Health Sciences, #42-121
Los Angeles, CA 90024
(213) 825-9111

Radiation physician well-known for his work at Chernobyl.

## General Federation of Women's Clubs
1734 N Street, NW
Washington, DC 20036
(202) 347-3168
*Alice C. Donahue, President*

Organizes volunteer service projects.

## Gibbs, Lois
P.O. Box 926
Arlington, VA 22216
(703) 276-7070

Love Canal activist, founded the Citizen's Clearinghouse for Hazardous Wastes in 1981.

## Green Seal
P.O. Box A.A., Stanford University
Palo Alto, CA 94605
(415) 327-2200
*Denis Hayes, Director*

National program which awards an environmental seal of approval on household products.

## Havens, Richie
10 East 44th Street, #700
New York, NY 10017

Folksinger and founder of the Natural Guard and the Northwind Undersea Institute, environmental education programs for urban children.

## Human Ecology Action League
P.O. Box 66637
Chicago, IL 60666
(312) 665-6575
*Ken Dominy, President*

Promotes awareness of environmental conditions that are hazardous to health and provides emotional support to victims of environment-related illnesses.

## Kelly, Petra K.
Die Grünen, Bundeshaus
HT 718, 5300 Bonn 1, Germany
0228-167918
0228-169206

Founder of die Grünen, the Green Party, in 1972 and one of the first political leaders in the environmental movement.

## Larkin, Hoffman, Daly and Lindgren, Ltd.
2000 Piper Jaffray Tower, 222 South North Street
Minneapolis, MN 55402
(612) 338-6610
*Richard A. Forschler, Vice-President*

Environmental law firm that also specializes in land use and development issues.

## Lovejoy, Tom
c/o Smithsonian Institution
1000 Jefferson Avenue, SW, Room 230
Washington, DC 20560
(202) 357-1300

Powerful lobbyist for environmental concerns.

## Maathai, Wangari
c/o Greenbelt Movement
P.O. Box 67545
Nairobi, Kenya

African environmental activist.

## National Trust for Historic Preservation
1785 Massachusetts Avenue, NW
Washington, DC 20036
(202) 673-4000
*Robert M. Bass, Chairman of the Board of Trustees*

Protects and preserves historical sites.

## Patagonia Environmental Program
P.O. Box 150
Ventura, CA 93002
(805) 643-8616

Manufacturer of outdoor clothing that donates 10 percent of its profits to hundreds of environmental groups.

## Richard King Mellon Foundation
525 William Penn Plaza
Pittsburgh, PA 15230
(412) 392-2800

Private foundation that donated more than 100,000 acres in 1990 for national parks, forests, and wildlife refuges—one of the biggest land contributions in U.S. history.

## Seeds of Change
621 Old Santa Fe Trail Road, #10
Santa Fe, NM 87501
(505) 983-8956
*Gabriel Howearth, Founder*

Encourages a plant-based diet to extend the earth's natural resources.

## Skinner, Nancy
2180 Milvia Street
Berkeley, CA 94704

Berkeley city council member who sponsored local law banning Styrofoam and initiated Stop Styro/Clearinghouse for Plastics and Packaging Reduction.

## StarKist Seafood Co.
180 E. Ocean Boulevard
Long Beach, CA 90802
(213) 590-9900
*Richard L. Beattie, CEO*

The first tuna company to switch to only buying tuna caught in dolphin-safe ways.

## Turner, Ted
P.O. Box 4064
Atlanta, GA 30302

Owner of Cable News Network, WTBS, and Turner Network Television. Broadcasts and produces many programs on the environment.

## U.S. Tourist Council
Drawer 1875
Washington, DC 20013-1875
(202) 293-1433
*Stanford West, Chairman*

Promotes eco-tourism—traveling in natural areas without disturbing the environment.

**Union of Concerned Scientists**
26 Church Street
Cambridge, MA 02238
(617) 547-5552
*Howard C. Ris, Jr., Executive Director*

Advocate organization concerned with the impact of advanced technology on society.

**Write For Action Group**
1339 61st Street
Emeryville, CA 94608
(415) 596-4040

Publishes EarthCards, postcards to help influence decision-makers on environmental issues.

# MAJOR ORGANIZATIONS

This chapter includes the most important, effective, and/or largest groups, international and national, involved with a wide range of issues in safeguarding our environment. Smaller organizations or those dealing mainly with a single subject may be found under specific subject headings.

## American Automobile Association (AAA)
8111 Gatehouse Road
Falls Church, VA 22047
(703) 222-6000
*Thomas McKernan, CEO*

With some 30 million members, this group is involved in many transportation issues, including fuel conservation and highway construction. Founded 1902.

## American Cancer Society
1599 Clifton Road
Atlanta, GA 30329
(404) 320-3333
*William M. Tipping, Executive Vice-President*

Environmentally speaking, involved in issues of air pollution, particularly from tobacco smoke, as well as other environmentally caused cancers.

## Center for Science in the Public Interest
1501 16th Street, NW
Washington, DC 20036
(202) 332-9110
*Michael F. Jacobson, Executive Director*

Consumer advocacy organization whose major interests are health and nutrition issues. Americans for Safe Food is a committee of the center.

## Concern, Inc.
1794 Columbia Road, NW
Washington, DC 20009
(202) 328-8160
*Susan Boyd, Executive Director*

Provides environmental information to individuals and groups and encourages action.

## Conservation Foundation, The
1250 24th Street, NW
Washington, DC 20037
(202) 293-4800
*Russell E. Train, Chairman of the Board*

Major research and public education organization founded in 1948.

## Earth Island Institute
300 Broadway, Suite 28
San Francisco, CA 94133
(415) 788-3666
*David Brower, Founder*

Major activist group. Includes the Save the Dolphins Project.

## Earthday
P.O. Box A.A., Stanford University
Palo Alto, CA 94605
(415) 321-1990
*Denis Hayes, Founder and Chairman*

Organized Earthday celebration and public awareness effort.

**Environmental Defense Fund**
257 Park Avenue South
New York, NY 10010
(212) 505-2100
(800) CALLEDF (Hotline)
*Frederic Krupp, Executive Director*

Organization of lawyers, scientists, and economists pursuing responsible reform of public policy in the fields of energy and resource conservation.

California
5655 College Avenue
Oakland, CA 94618
(415) 658-8008

Colorado
1405 Arapahoe
Boulder, CO 80302
(303) 440-4901

North Carolina
128 East Hargett Street
Raleigh, NC 27601
(919) 821-7793

Virginia
1108 E. Main Street
Richmond, VA 23219
(804) 780-1297

Washington
1616 P Street, NW
Washington, DC 20036
(202) 387-3500

**Food and Agriculture Organization of the United Nations (FAO)**
Via delle Terme di Carcalla
Rome 00100, Italy
06 57971
*Edouard Saouma, Director-General*

U.N. agency concerned with nutrition and the distribution of food and agricultural goods, including forestry products.

**4-H Program**
Extension Service
U.S. Department of Agriculture
Washington, DC 20250
(202) 447-5853
*Leah Hooper, Deputy Administrator*

The youth education program of the Cooperative Extension Service, with more than four million members. Founded 1900.

**Friends of the Earth**

Promotes conservation, restoration, and rational use of the environment and the earth's natural resources through public education and campaigning at the local, national, and international levels.

Argentina (Amigos de la Tierra)
Anchorena 633
1170 Buenos Aires, Argentina
88-3815

Australia
P.O. Box 530-E
Melbourne, Victoria 3001
Australia

Austria
Rembrandtstrasse 14
A-1020 Vienna, Austria

Belgium (Les Amis de la Terre)
1 rue de l'Esplanade
B-1050 Brussels, Belgium

Brazil (Amigos da Terra)
Rua Miguel Tostes 694
Porto Alegre 90.000, Brazil

Canada
701-251 Laurier Avenue West
Ottawa, Ontario K1P 5J6
Canada

Cyprus
Lanarca District
Maroni, Cyprus

Ecuador (Tierra Viva)
Casilla 1891
Cuenca, Ecuador

El Salvador (Amigos de la Tierra
de El Salvador)
Segunda Avenid Norte 1-2
Santa Tecla, San Salvador, El
Salvador
25-2603

England
377 City Road
London EC1V 1NA England
837 0731

France (Les Amis de la Terre)
15 rue Gambay
F-75011, Paris, France

Ghana
P.O. Box 3794
Accra, Ghana

Hong Kong
One Earth Centre, 61 Wyndham
Street 1/F, Mezzanine Floor
Central Hong Kong

Ireland (Earthwatch)
Harbour View, Bantry
County Cork, Ireland

Italy (Amici della Terra)
Piazza Sforza Cesarini 28
I-00186 Rome, Italy

Japan (Chikyu no Tomo)
501 Shinwa Building, 9-17
Sakuragaoka,
Shibuya-ku, Tokyo 150, Japan

Malaysia (Sahabat Alam)
43 Salween Road
10050 Penang, Malaysia
376930

New Zealand
P.O. Box 39-065
Auckland West, New Zealand
34-319

Papua New Guinea
P.O. Box 4028
Boroko, New Guinea

Portugal (Amigos da Terra)
rue Pinheiro Chaves 28
2 Dto., P-1000 Lisbon, Portugal

Scotland
53 George IV Bridge
Edinburgh EH1 1EJ, United
Kingdom
225 6906

Spain (Federacion de Amigos de
la Tierra)
Avenida Betanzos 55, 11.1
E-28025 Madrid, Spain

Sweden (Jordens Vanner)
Regeringsgatan 70C
S-111 39 Stockholm, Sweden

## Friends of the Earth, Environmental Policy Institute, The Oceanic Society

Executive Offices, 218 D Street, SE
Washington, DC 20003
(202) 544-2600
*Michael Clark, President*

The merger of these three environmental groups in 1990 continues to work to empower citizens to effect change.

Long Island Sound Task Force
Stamford Marine Center, 185
Magee Avenue
Stamford, CT 06902
(203) 327-9786
*Kathryn Clarke, President*

Los Angeles Chapter
1415 3rd Street, Suite 300A
Santa Monica, CA 90401

San Francisco Bay Chapter
Fort Mason Center, Bldg. E
San Francisco, CA 94123
(415) 441-5970
*Margaret Elliot, Executive
Director*

## Friends of the United Nations Environment Programme (FUNEP)

2013 Q Street, NW
Washington, DC 20009
(202) 234-3600
*Richard A. Hellman, Executive
Director*

Links UNEP to environmental groups in the U.S.

## Global Tomorrow Coalition
1325 G Street, NW, Suite 915
Washington, DC 20005-3104
(202) 628-4016
*Donald R. Lesh, President*

Coalition of 115 organizations, with 8,000,000 members. Emphasizes global concern over environment and resources.

## Greenpeace Action
**Organization of local groups. Also see Greenpeace International and Greenpeace USA.**

Amherst
253-A Triangle Street
Amherst, MA 01002
(413) 549-0507
*Bill Richardson, Representative*
*Diane Milano, Representative*

Ann Arbor
214 North 4th Street
Ann Arbor, MI 48104
(313) 761-1996
*Jeff Muhr, Representative*
*Chris Law, Representative*

Atlanta
20 13th Street, NE
Atlanta, GA 30303
(404) 874-7585
*Don Kelly, Representative*
*Amy Conly, Representative*

Austin
1403 Rio Grande
Austin, TX 78701
(512) 549-0507
*Janet Manley, Representative*
*John Lofy, Representative*

Boston
709 Center Street
Jamaica Plain, MA 02130
(617) 983-0300
(617) 983-0909 FAX
*Ingrid Gordon, Representative*

Boulder
2025 16th Street
Boulder, CO 80302
(303) 440-3381
*Suzanne Pomeroy,*
*Representative*
*Jack Mento, Representative*

Chicago
1017 West Jackson
Chicago, IL 60607
(312) 666-3305
*Scott Mendrick, Representative*
*Jane Gire, Representative*

Cincinnati
2826 Euclid Avenue
Cincinnati, OH 45219
(513) 281-4242
*Orson Moon, Representative*
*Steve Backs, Representative*

Ft. Lauderdale
400 South Andrews Street
Ft. Lauderdale, FL 33301
(305) 768-9660
(305) 768-0711 FAX
*Stephanie Settler, Representative*

Kansas City
1613 West 39 Street
Kansas City, MO 64111
(316) 531-3884
*Steve Land, Representative*
*Amy Hadan, Representative*

Los Angeles
8599 Venice Boulevard
Los Angeles, CA 90034
(213) 287-2210
(213) 287-0832 FAX
*Lynn Howard, Representative*

Madison
1053 East Williamson Street
Madison, WI 53703
(608) 251-2661
*Jim Naresore, Representative*
*Mairane Ruppel, Representative*

Minneapolis
2637 Nicollet Avenue, South
Minneapolis, MN 55408
(612) 974-0320
*Bill Galloway, Representative*
*Bill Busse, Representative*

Nashville
1025 17th Avenue South,
  Suite D
Nashville, TN 37212
(615) 327-7995
*Theresa Duren, Representative*
*Leslie Harris, Representative*

New Haven
334 Whalley Avenue, #328
New Haven, CT 06511
(203) 785-0198
*Mark Schuyler, Representative*
*Andy Gresh, Representative*

New York
96 Spring Street, 3rd Floor
New York, NY 10012
(212) 941-0994
*Gordon White, Representative*
*Helen Gardiner, Representative*

Orlando
831 North Irea Street
Orlando, FL 32805
(407) 648-8222
*Gary English, Representative*
*Rob O'Brien, Representative*

Philadelphia
785 South 6th Street
Philadelphia, PA 19147
(215) 925-2075
*Kathy Armstrong, Representative*
*Andrew Altman, Representative*

Phone Bank
709 Centre Street
Jamaica Plain, MA 02130
(800) 456-4029 Catalog
(800) 327-3222 Customer Service
*Barry Pike, Spokesperson*

Portland
1437 S.W. Columbia
Portland, OR 97201
(503) 241-1507
*Joe Keating, Representative*
*Perry Miller, Representative*

Raleigh
723 Johnson Street
Raleigh, NC 27603
(919) 934-6595
*Lauren Ivey, Representative*
*Frank Lucca, Representative*

Rochester
302 North Goodman Street,
  #409-410
Rochester, NY 14607
(716) 442-0870
*Kelly Wilkens, Representative*
*Miriam Steinberg, Representative*

San Diego
3909 4th Avenue, #201
San Diego, CA 92103
(619) 293-1010
*Maureen Burkett, Representative*

San Francisco
Fort Mason Center, Building E
San Francisco, CA 94123
(415) 474-6767
(415) 346-6110
*Ray Gish, Representative*

Santa Cruz
1112 B Ocean Street
Santa Cruz, CA 95060
(408) 429-9988
*Richard Jorgensen,*
*Representative*

Seattle
3420 Stone Way Avenue, N
Seattle, WA 98103
(206) 633-6027
*Richard Green, Representative*
*Eric Glynn, Representative*

Washington, DC
1638 R Street, NW
Washington, DC 20009
(202) 667-7914
*Mike Phillips, Representative*
*S. Thompson, Representative*

**Greenpeace International**
Temple House, 25-26 High Street
Lewes, East Sussex BN7 2LU
England
478787
471631 FAX
*David McTaggart, Executive
Director*

Believes in using direct action in its campaigns to protect wildlife and the environment. Also see Greenpeace Action and Greenpeace USA.

Argentina
Junin 45, 3rd Piso
1026 Buenos Aires, Argentina
1 953-3336
1 311-3111 FAX
*Melvyn Gattinoni, Representative*

Australia
155 Pirie Street
Adelaide, South Australia 5000,
Australia
8 223-3578
8 211-7720 FAX
*Cheryl McEgan, Representative*

Austria
Mariahilfer Guertel 32
A-1060 Vienna, Austria
222 597-3046
222 597-3055 FAX
*Wladimir Zalozieckyj,
Representative*

Belgium
Waversesteenweg 335
B-1040 Brussels, Belgium
2 647-8765
2 647-8782
*Ward Dossche, Representative*

Canada
5798 Bloor Street West
Toronto, Ontario M6G 1K1
Canada
(416) 538-6470
(416) 538-6479 FAX
*Michael Manolson,
Representative*

Communications International
123 Cannon Workshops, West
India Dock
London E14 England
44 1 515-0275
*Martin Leeburn, Greenpeace
Contact
Sara Saunders, Photo Division
Liz Grey, Film Division*

Denmark
Thomas Laubsgade 11-13,
DK-2100
Copenhagen O, Denmark
1 185-444
1 184-267 FAX
*Michael Neilsen, Representative*

Germany
Vorsetzen 53
D-2000 Hamburg 11, Germany
49 40 311-860
49 40 311-86-141 FAX
*Gerhard Wallmeyer,
Representative*

Ireland
29 Lower Baggot Street
Dublin 2, Ireland
619-836
610-814 FAX
*John Bowler, Representative*

Italy
28 Viale Manlio Gelsomini
00153 Rome, Italy
39 6 578-2486
39 6 578-3531 FAX
*Gianni Squitieri, Representative*

Latin American Regional Office
4th Piso, Edifico Noga
Frente Centro Colon, Paseo
Colon, Costa Rica
571-005
571-008 FAX
*Dorothy Houston, Representative*

Luxembourg
P.O. Box 229
4003 Esch/Alzette, Luxembourg
352 546-252
*Roger Spautz, Representative*

Netherlands
Damrak 83
1012 LN Amsterdam
The Netherlands
31 20 261-877
31 20 221-272 FAX
*Ron Van Huizen, Representative*

New Zealand
Private Bag, Wellesley Street
Auckland, New Zealand
64 9 776-128
64 9 32676 FAX
*Bev Cormack, Representative*

Norway
St. Olavsgt 11
P.O. Box 6803 St. Olavaplass
0130 Oslo 1, Norway
47 2 205-101
47 2 205-114 FAX
*Erik Robsahm, Representative*

Spain
Rodriguez San Pedro 58, 4 Piso
28015 Madrid, Spain
34 1 243-4704
34 1 243-9779
*Kiki Van Lochen, Representative*

Sweden
Box 7629
103 94 Stockholm, Sweden
46 8 118-828
46 8 247-468
*Toruny Ekstrom, Representative*

Switzerland
Meullarstrasse 37, Postfach 4927
8004 Zurich, Switzerland
41 1 241-3441
41 1 241-3821 FAX

United Kingdom
30–31 Islington Green
London N1 8XE England
44 1 354-5100
44 1 359-4062 FAX
*Mary Morrison, Representative*

**Greenpeace USA**
1436 U Street, NW
Washington, DC 20009
(202) 462-1177
(800) 333-7717
*Peter Bahouth, Executive Director*

National leadership. Also see Greenpeace Action and Greenpeace International.

**International Union For Conservation of Nature and Natural Resources (IUCN)**
Avenue du Mont-Blanc
CH-1196 Gland, Switzerland
022.64 91 14
*Monkombu S. Swaminathan, President*

Independent organization of nations, government agencies, and citizen's groups promoting scientific action for the preservation of the environment.

**National Audubon Society**
950 Third Avenue
New York, NY 10022
(212) 832-3200
*Harold E. Woodsum, Jr., Chairman of the Board*

Founded in 1905, more than 500,000 members support broad conservation activities.

Alaska
308 "G" Street, Suite 217
Anchorage, AK 99501
(907) 276-7034
*David Cline, Vice-President*

Education Division Headquarters
Rt. 1, Box 171
Sharon, CT 06069
(203) 364-0520
*Marshal T. Case, Vice-President*

Field Research and
Ornithological Research
Headquarters
115 Indian Mound Trail
Tavernier, FL 33070
(305) 852-5092
*Alexander Sprunt IV,*
*Vice-President*

Great Lakes (IL, IN, MI, MN,
OH, WI)
7 North Meridian Street, #400
Indianapolis, IN 46204
(317) 631-2676
*David Newhouse, Vice-President*

Hawaii
212 Merchant Street, Suite 320
Honolulu, HI 96813
(808) 522-5566
*Dana Kokubun, State*
*Representative*

Mid-Atlantic (DE, MD, NJ, PA,
VA, WV)
1104 Fernwood Avenue, Suite
300
Camp Hill, PA 17011
(717) 763-4985
*Walt Pomerory, Vice-President*

Northeast (CT, ME, MA, NH,
NY, RI, VT)
1789 Western Avenue
Albany, NY 12203
(518) 869-9731
*David Miller, Vice-President*

Rocky Mountain (AZ, CO, ID,
MT, UT, WY)
4150 Darley, Suite 5
Boulder, CO 80303
(303) 499-0219
*Robert Turner, Vice-President*

Sanctuaries
R.R. 1, Box 294
Sharon, CT 06069
(203) 364-0048
*Frank M. Dunstan,*
*Vice-President and Director*

Southeast (AL, FL, GA, KY,
MS, NC, SC, TN)
928 North Monroe Street
Tallahassee, FL 32303
(904) 222-2473
*Larry Thompson, Vice-President*

Southwest (LA, NM, TX,
Guatemala, Mexico, Panama)
2525 Wallingwood, #1505
Austin, TX 78746
(512) 327-1943
*Dede Armentrout, Vice-President*

Washington, DC
801 Pennsylvania Avenue, SE,
Suite 301
Washington, DC 20003
(202) 547-9009
*Brooks Yeager, Vice-President*

West Central (AR, IA, KS, MO,
NE, ND, OK, SD)
200 Southwind Place, #205
Manhattan, KS 66502
(913) 537-4385
*Ron Klataske, Vice-President*

Western (CA, NV, OR, WA,
Guam)
555 Audubon Place
Sacramento, CA 95825
(916) 481-5332
*Glenn Olson, Vice-President*

**National Geographic Society**
1145 17th Street (at M Street)
Washington, DC 20036
(202) 857-7000
*Gilbert Grosvenor, Chairman of the*
*Board*

Publishes *National Geographic* magazine, produces television specials, funds
research, and educates the public on
earth issues.

**National Wildlife Federation**
1400 Sixteenth Street, NW
Washington, DC 20036-2266
(202) 797-6800
*George H. Hulsey, Chairman of the*
*Board*

Affiliation of more than 5,000,000 members emphasizing public education on wildlife issues.

Region 1 (CT, ME, MA, NH, RI, VT)
Hollow Road
North Ferrisburgh, VT 05473
(802) 425-2329
*Carl Reidel, Regional Director*

Region 2 (DE, MD, NJ, NY, PA)
Box 267
Millerstown, PA 17062
(717) 589-3929
*Ralph W. Abele, Regional Director*

Region 3 (NC, SC, VA, WV)
P.O. Box 527
Ellerbe, NC 28338
(919) 652-5061
*John F. Lentz, Regional Director*

Region 4 (AL, FL, GA, MS, PR, VI)
Rt. 1, Box 50
Alma, GA 31510
(912) 383-4356
*Delano Deen, Regional Director*

Region 5 (AR, KY, MO, TN)
3800 Capitol Tower Building
Little Rock, AR 72201
(501) 375-9151
*Don F. Hamilton, Regional Director*

Region 6 (IL, IN, OH)
R.R. 1, Box 465
Parker City, IN 47368
(317) 468-7410
*Emily M. Kress, Regional Director*

Region 7 (MI, MN, WI)
735 Crystal Lake Road
Burnsville, MN 55337
(612) 774-6600
*Gordon G. Meyer, Regional Director*

Region 8 (LA, OK, TX)
720 Brazos, Suite 900
Austin, TX 78701
(512) 476-7028
*Tom Martine, Regional Director*

Region 9 (IA, KS, NE, ND, SD)
1319 4th Street
Brookings, SD 57006
(605) 692-6748
*Raymond L. Linder, Regional Director*

Region 10 (AZ, CO, NM, VT)
Box 13938
Ft. Carson, CO 80913
(719) 579-4828
*Thomas L. Warren, Regional Director*

Region 11 (AK, OR, WA)
c/o NWF AK Nat. Res. Ctr.,
750 West Second Ave.,
Suite 200
Anchorage, AK 99501
(907) 562-3366
*James E. Hemmings, Regional Director*

Region 12 (CA, HI, NV)
2820 Echo Way
Sacramento, CA 95821
(916) 971-1953
*Rudy J.H. Schafer, Regional Director*

Region 13 (ID, MT, WY)
714 Sun Valley Drive
Cheyenne, WY 82001
(307) 777-6564
*Dan E. Cunningham, Regional Director*

## Natural Resources Council of America
1015 31st Street, NW
Washington, DC 20007
(202) 333-8495
*Paul C. Pritchard, Chairman*

Association of major national and regional organizations concerned with the sound management of natural resources in the public interest.

**Natural Resources Defense Council**
40 West 20th Street
New York, NY 10011
(212) 727-2700
*John H. Adams, Executive Director*

Protects endangered natural resources.

Hawaii
212 Merchant Street, Suite 203
Honolulu, HI 96813
(808) 533-1075

Los Angeles
617 South Olive Street
Suite 1210
Los Angeles, CA 90017
(213) 892-1500

San Francisco
90 New Montgomery
San Francisco, CA 94105
(415) 777-0220

Washington
1350 New York Avenue, NW
Washington, DC 20005
(202) 783-7800

**Nature Conservancy, The**
1815 North Lynn Street
Arlington, VA 22209
(703) 841-5300
*David L. Harrison, Chairman of the Board*

Committed to preserving ecological diversity through the protection of natural areas. Founded 1917.

Eastern Regional Office
294 Washington Street,
Room 740
Boston, MA 02108
(617) 542-1908
*Dennis B. Wolkoff,
Vice-President*

Hawaii
1116 Smith Street, #201
Honolulu, HI 96817

Midwest Regional Office
1313 5th Street, SE
Minneapolis, MN 55414
(612) 379-2207
*Russ Van Herik, Vice-President*

Southeast Regional Office
P.O. Box 270
Chapel Hill, NC 27514
(919) 967-5493
*Chuck Bassett, Vice-President*

Western Regional Office
785 Market Street, 3rd Floor
San Francisco, CA 94103
(415) 777-0541
*Laurel Mayer, Vice-President*

**Public Citizen**
P.O. Box 19404
Washington, DC 20036
(202) 293-9142
*Joan Claybrook, President*
*Ralph Nader, Founder*

Supports the work of citizen advocates for consumer rights, a healthful environment, safe energy sources, and corporate and government accountability.

**Sierra Club**
730 Polk Street
San Francisco, CA 94109
(415) 776-2211
*Richard Cellarius, President*

Founded in 1892, one of the largest and most well-known environmental organizations in the world. Broad-ranging concerns and projects.

Alaska Representatives
241 East 5th Avenue, #205
Anchorage, AK 99501
(907) 276-4048
*Jack Hesson, Representative*
*Emily Barnett, Representative*

Appalachian Representatives
1116 #C Street
Annapolis, MD 21404-0667
(301) 268-7411
*Ken Gersten, Representative*
*Joy Oakes, Representative*

Canada—Ontario
229 College Street, Room 303
Toronto, Ontario M5T 1R4
Canada
(416) 596-7778

Canada—Western Canada
620 View Street, Room 314
Victoria, British Columbia V8W
1J6 Canada
(604) 386-5255

Midwest Representatives
214 North Henry Street,
Suite 203
Madison, WI 53703
(608) 257-4994
*Jane Elder, Representative*
*Leanne Klyza-Linck,*
*Representative*

Northeast Representative
360 Broadway
Saratoga Springs, NY 12866
(518) 587-9166
*Chris Ballantyne, Representative*

Northern California–Nevada
Representatives
5428 College Avenue
Oakland, CA 94618
(415) 654-7847
*Sally Kabisch, Representative*

Northern Plains Representative
23 North Scott, Room 25
Sheridan, WY 82801
(307) 672-0425
*Larry Mehlhaff, Representative*

Northwest Representative
1516 Melrose Avenue
Seattle, WA 98122
(206) 621-1696
*Bill Arthur, Representative*
*Barbara Boyle, Representative*

Southeast Representative
(Florida)
1201 North Federal Highway,
Room 250-H
North Palm Beach, FL 33408
(407) 775-3846
*Theresa Woody, Representative*

Southeast Representative
(Tennessee)
P.O. Box 11248
Knoxville, TN 37939-1248
(615) 588-1892
*James Price, Representative*

Southern California-Nevada
Representatives
3550 West 6th Street, #323
Los Angeles, CA 90020
(213) 387-6528
*Bob Hattoy, Representative*
*Jeff Widen, Representative*

Southern Plains Representative
7502 Greenville Avenue, #170
Dallas, TX 75231
(214) 824-5930
*Beth Johnson, Representative*

Southwest Representative
(Arizona)
3201 North 16th Street, #6-A
Phoenix, AZ 85016
(602) 277-8079
*Rob Smith, Representative*

Southwest Representative
(Colorado)
1240 Pine Street
Boulder, CO 80302
(303) 449-5595
*Maggie Fox, Representative*

Southwest Representative (Utah)
177 East 900 South, #102
Salt Lake City, UT 84111
(801) 355-0509
*Lawson Legate, Representative*

Washington, DC
408 C Street, NE
Washington, DC 20002
(202) 547-1144
*David Gardiner, Legislative*
*Director*

## Sierra Club Legal Defense Fund
2044 Fillmore Street
San Francisco, CA 94115
(415) 567-6100
*Fredric P. Sutherland, Executive*
*Director*

Supports lawsuits brought on behalf of citizens' organizations for environmental action.

Alaska
325 4th Street
Juneau, AK 99801
(907) 586-2751
*Lauri J. Adams, Staff Attorney*
*Eric P. Jorgensen, Staff Attorney*

Hawaii
212 Merchant Street, Arcade Building
Honolulu, HI 96813
(808) 599-2436
*Arnold Lum, Staff Attorney*

Rocky Mountain Office
1600 Broadway Street, Suite 1600
Denver, CO 80202
(303) 863-9898
*Lori Potter, Staff Attorney*
*Mark Hughes, Staff Attorney*

Seattle
216 1st Avenue, Suite 330
Seattle, WA 98104
(206) 343-7340
*Victor Sher, Staff Attorney*
*Todd True, Staff Attorney*

Washington, DC
1531 P Street, NW, Suite 200
Washington, DC 20005
(202) 667-4500
*Ronald J. Wilson, Counsel*

## Smithsonian Institution
1000 Jefferson Drive, SW
Washington, DC 20560
(202) 357-1300
*Robert McCormick Adams, Secretary*

Established in 1846 "for the increase and diffusion of knowledge among men." A major funding and research umbrella for environmental studies.

## Smithsonian Environmental Research Center
P.O. Box 28
Edgewater, MD 21037
(301) 261-4190
*David L. Correll, Director*

## U.S. Chamber of Commerce
1615 H Street, NW
Washington, DC 20062
(202) 659-6000
*Richard L. Lesher, President*

National federation of business organizations and companies. Has committees on energy, environment, and natural resources. Founded in 1912.

## United Nations Environment Programme (UNEP)

Headquarters and Regional Office for Africa
P.O. Box 30552
Nairobi, Kenya
333930
520711 FAX
*Mostafa Kamal Tolba, Executive Director*
Founded in 1972. Coordinates and stimulates environmental action largely within the U.N. system but also the international community.

Co-ordinating Unit for the Mediterranean Action Plan
Leoforos Vassileos Konstantinou 48
Athens 501/1, Greece
72 44 536

New York Liaison Office
Room DC2-0816
New York, NY 10017
(212) 963-8138

Regional Co-ordination Unit (Caribbean)
14-21 Port Royal Street
Kingston, Jamaica
92 29 269

Regional Office for Asia and the Pacific
ESCAP, Rajadamnern Avenue
Bangkok 10200, Thailand
829161-200

Regional Office for Latin America
Edificio Naciones Unidas, Presidente Mazaryk 29
Apartado Postal 6-718, Mexico City 5, D.F., Mexico
2501555

Regional Office for West Asia
P.O. Box 10880
Manama, Bahrain
27 60 72

Regional Office for Europe
Pavillons du Petit-Saconex, 16 avenue Jean Trembley
CH-1209 Geneva, Switzerland
798 84 00

Washington Liaison Office
1889 F Street, NW
Washington, DC 20006
(202) 289-8456

## United Nations Secretariat for International Trade in Endangered Species of Wild Fauna and Flora

6 rue du Maupas, Case Postale 78
CH-1000 Lausanne 9, Switzerland
20 00 81

## World Health Organization

Avenue Appia CH-1211
Geneva 27, Switzerland
*Hiroshi Nakajima, General Director*

Agency of the United Nations established in 1948 to further international cooperation for improved health conditions around the world.

## Worldwatch Institute

1776 Massachusetts Avenue, NW
Washington, DC 20036
(202) 452-1999
*Lester Brown, Founder*

Publishes *State of the World Report,* focusing on ozone, food production, and alternative energy.

# AIR

**Problem:** From acid rain to smog, our air is being polluted. From the greenhouse effect to depletion of the ozone layer, our atmosphere is being destroyed.

**Solution:** Curbing the use of fossil fuels, such as coal, oil, and natural gas, in industrial plants and motor vehicles; eliminating CFCs (chlorofluorocarbons); and planting rather than burning trees, will significantly improve the quality of our air and climate for human, animal, and plant life.

This chapter includes government and non-government groups, international agencies, organizations, companies, and individuals mainly involved with: acid rain, air pollution, atmospheric studies, automobiles, chlorofluorocarbons, ozone issues, radon, tobacco smoke, and transportation. (For fossil fuel producers, see *Energy.* For forest issues, see *Land.*)

## Acid Precipitation Data Base (ACIDOC)

N.Y. State Dept. of
Environment—Div. of Air
Resources
50 Wolf Road
Albany, NY 12233
(518) 457-2823

International bibliography on acid rain.

## Acid Rain Foundation

1410 Varsity Drive
Raleigh, NC 27606
(919) 828-9443
*Dr. Rodger W. Bybee, Director*
*Dr. Ellis B. Cowling, Director*

Strives to bring about a greater understanding of the acid rain problem.

## Air and Waste Management Association

P.O. Box 2861
Pittsburgh, PA 15230
(412) 232-3444
*Charles D. Pratt, President*

Formerly Air Pollution Control Association, which was founded in 1907. Publishes *The International Journal of Air Pollution Control and Hazardous Waste Management.*

## Air Chek

180 Glenn Bridge Road, Box 2000
Arden, NC 28704
(800) AIR-CHEK

Manufactures airborne radon test equipment for the home.

## Air Pollution Action Network

Postbus 5627
N-1007 AP Amsterdam,
Netherlands

International organization formed in 1985.

## Air Resources Information Clearinghouse (ARIC)

Center for Environmental Information
99 Court Street
Rochester, NY 14604
(716) 546-3796
*Elizabeth Thorndike, President*
*Frederick W. Stoss, Director of Library and Information Services*

Comprehensive reference and referral educational service on acid rain, the greenhouse effect, ozone depletion, etc.

## Alliance for Clean Energy

1901 North Ft. Meyer Drive, 12th Floor
Arlington, VA 22209
(703) 841-1781
*Nancy Prowitt, Assistant Director*

An alliance of low-sulphur coal producers, users, and transporters interested in the acid rain problem.

## Alliance for Responsible CFC Policy

1901 North Ft. Meyer Drive
Arlington, VA 22209
(703) 841-9363
*Richard C. Barnett, Chairman*

Group composed of users and producers of chlorofluorocarbon chemicals.

## American Brands

245 Park Avenue
New York, NY 10167
(212) 880-4200
*Edward W. Whittemore, CEO*

Manufactures tobacco products.

## American Lung Association

1740 Broadway
New York, NY 10019-4374
(212) 315-8700
*John R. Garrison, Managing Director*

Involved in many aspects of air pollution control, particularly tobacco use. Established 1904.

## American Pedestrian Association

P.O. Box 624, Forest Hills Station
Forest Hills, NY 11375
*M. Carasso, President*

Lobbies against vehicular encroachment. First pedestrian group in the U.S.

## American Public Transit Association

1201 New York Avenue, NW, Suite 400
Washington, DC 20005
(202) 898-4000
*Jack R. Gilstrap, Executive Vice-President*

Organization of rapid rail and motor bus transit systems in the U.S., Canada, and Mexico.

## Amtrak—National Railroad Passenger Corporation

400 North Capitol Street, NW
Washington, DC 20001
(202) 383-3000
*W. Graham Clayton, Jr., Chairman*

Intercity rail passenger service.

## Appliance Recycling Centers of America

654 University Avenue
St. Paul, MN 55104
(612) 291-1100

Recycles the Freon in refrigeration and air-conditioning units.

## Association for Commuter Transportation

1776 Massachusetts Avenue, NW, #521
Washington, DC 20036
(202) 659-0602
*Sandra Spence, Executive Director*

Promotes expansion of ride-sharing, van pools, and other alternative commuting modes.

## Association of Local Air Pollution Control Officials

444 North Capitol Street, NW, Suite 306
Washington, DC 20001
(202) 624-7864
*S. William Becker, Executive Director*

Publishes *Washington Update,* a newsletter covering congressional and EPA activities.

## Bicycle Federation of America

1818 R Street, NW
Washington, DC 20009
(202) 332-6986
*William C. Wilkinson, Executive Director*

Promotes bicycling for transportation and recreation.

## Bicycle Network

P.O. Box 8194
Philadelphia, PA 19101
(215) 222-1253
*John Dowlin, Editor*

Supports bicycling as a healthful and energy-efficient mode of transportation.

## Bowker A&I Publishing

245 West 17th Street
New York, NY 10017

Publishes *Acid Rain Abstracts Annual,* a bi-monthly publication.

## California Air Resources Board

P.O. Box 2815
Sacramento, CA 95812
(916) 322-2990
*Jananne Sharpless, Chairwoman*

## Canadian Coalition on Acid Rain

401–112 St. Clair Avenue West
Toronto, Ontario M4V 2Y3
Canada
(416) 968-2135

## Chrysler Motors

12000 Chrysler Drive
Highland Park, MI 48288
(313) 956-5252
*Lee Iacocca, Chairman*

## Citizens Against Tobacco Smoke

P.O. Box 36236
Cincinnati, OH 45236
(513) 984-8834
*Ahron Leichtman, President*

Individuals, health organizations, and environmental groups concerned with indoor pollution caused by tobacco smoke.

## Citizens for Sensible Control of Acid Rain

1301 Connecticut Avenue, NW, Suite 700
Washington, DC 20036
(202) 659-0330
*Thomas L. Buckmaster, Executive Director*

Purpose is to educate the public on clean air policy initiatives. Funded by electric, coal, and manufacturing companies.

## Clean Air Working Group

818 Connecticut Avenue, NW, #900
Washington, DC 20006
(202) 857-0370
*William D. Fay, Administrator*

Promotes clean air through individuals taking part in ecological practices on a day-to-day basis.

## Climate Institute
316 Pennsylvania Avenue,
Suite 403
Washington, DC 20003
(202) 547-0104

Researches global climate change.

## Coalition for Clean Air
309 Santa Monica Boulevard,
Suite 212
Santa Monica, CA 90401
(213) 451-0651
*Jan Chatten-Brown, President*

Dedicated to the eradication of air pollution in California.

## Cyrodynamics
1101 Bristol Road
Mountainside, NJ 07092

Has developed air-conditioners and refrigerators that do not use CFCs.

## Earth Regeneration Society
1442A Walnut Street, #57A
Berkeley, CA 94709
(415) 525-7723
*Alden Bryant, President*

An organization concerned primarily with the greenhouse effect.

## Ford Motor Company
The American Road
Dearborn, MI 48121
(313) 322-3000
*Donald E. Petersen, Chairman*

## General Motors Corporation
3044 West Grand Boulevard
Detroit, MI 48202
(313) 248-6100
*Robert Stemple, CEO*

## Greenhouse Crisis Foundation
1130 17th Street, NW, Suite 630
Washington, DC 20036
(202) 466-2823
*Jeremy Rifkin, Founder*

## Group Against Smokers' Pollution (GASP)
P.O. Box 632
College Park, MD 20740
(301) 577-6427
*Willard K. Morris, Secretary*

Promotes non-smokers' rights against second-hand smoke.

## Hansen, James
Goddard Institute for Space
Studies
2880 Broadway
New York, NY 10025
(212) 678-5500

Global warming expert.

## Honda
2-6-20 Yaesu, Chu-ku
Tokyo 104, Japan
*Soichiro Honda, Chairman*

## Institute for Transportation and Development Policy
P.O. Box 56538
Washington, DC 20011
(301) 589-1810
*Michael A. Replogle, President*

Creators of Bikes Not Bombs and Bikes for Africa programs. Promotes ecologically sound transportation in the Third World.

## International Bicycle Fund
4247 135th Place, S.E.
Bellevue, WA 98006
(206) 746-1028
*David Mozer, Director*

Promotes transportation by bicycle.

## International Human Powered Vehicle Association
P.O. Box 51255
Indianapolis, IN 46251
(317) 876-9478
*Marti Daily, President*

Engineers, academic researchers, bicycling enthusiasts and others interested in human-powered vehicles.

## Japan Air Cleaning Association
Tomoeya Building, 2-14 Uchikanda 1-chome
Chiyoda-ku, Tokyo 101, Japan
233-1486

## League of American Wheelmen
6707 Whitestone Road, Suite 209
Baltimore, MD 21207
(301) 944-3399
*John M. Cornelison, Administrator*

Has campaigned for bicycle riding for more than 100 years.

## Manufacturers of Emission Controls Association
1707 L Street, NW, Suite 570
Washington, DC 20036
(202) 296-4797
*Bruce I. Bertelsen, Executive Director*

Manufacturers of automobile exhaust control devices and stationary-source catalytic controls.

## Mazda Motor Corporation
3-1 Sinchi, Fuchu-ocho, Akigun
Hiroshima 730-19 Japan
*Kenichi Yamamoto, Chairman*

## Mobile Air Conditioning Society
7425 Westchester Pike
Upper Darby, PA 19082
(215) 352-1345

Updates list of businesses that recycle refrigerants from automobiles.

## National Aeronautics and Space Administration (NASA)
600 Independence Avenue, SW
Washington, DC 20546
(202) 453-1000
*Richard Truly, Administrator*

Studies ozone, global warming, and other atmospheric conditions.

## National Association of Transit Consumer Organizations
442 Summit Avenue, #2
St. Paul, MN 55102
(612) 227-5171
*Corbin S. Kidder, Executive Director*

Primary goal is to maintain and improve urban transit services.

## National Center for Atmospheric Research
P.O. Box 3000
Boulder, CO 80307
(303) 494-5151
*Erich Bloch, Director*

Sponsors research on global climate change.

## National Clean Air Coalition
1400 16th Street, NW
Washington, DC 20036
(202) 797-5436
*Susan Buffone, Executive Director*

Works to strengthen the Clean Air Act.

## Philip Morris Companies
120 Park Avenue
New York, NY 10017
(212) 880-5000
*Hamish Maxwell, CEO*

Producer of Marlboro, largest selling cigarette in the world.

## RJR Nabisco
Post Office Box 2959
Winston-Salem, NC 27102
(919) 773-2000
*J. Tylee Wilson, Chairman of the Board*

Produces tobacco products.

## Rowland, F. Sherwood
University of California
Irvine, CA 92717
(714) 856-5190

Scientist who discovered link between chlorofluorocarbons and the destruction of the ozone layer in 1973.

## Sagan, Carl Edward
c/o Space Sciences Building
Cornell University
Ithaca, NY 14853
(607) 255-2000

Space scientist, concerned about the ozone layer and global warming.

## Skin Cancer Foundation
245 Fifth Avenue, Suite 2402
New York, NY 10016
(212) 725-5176
*Mitzi Moulds, Executive Director*

Sponsors medical symposia and public education programs on the prevention and early recognition of skin cancer.

## South Coast Air Quality Management District
9150 Flare Drive
El Monte, CA 91731
(800) 242-4022 (Air Quality)
(800) 242-4666
*James Lentz, Executive Officer*

Monitors air quality throughout southern California.

## Taylor-Dunn Manufacturing
2114 West Ball Road
Anaheim, CA 91804
(714) 956-4041
*R. Davis Taylor II, President*

Manufacturer of electric vehicles since 1949.

## ThermaFlo
3640 Main Street
Springfield, MA 01107
(800) 556-6015

Markets a CFC reclamation unit for refrigeration systems.

## Tobacco Institute
1875 I Street, NW, Suite 800
Washington, DC 20006
(202) 457-5151
*Samuel Chilcote, President*

Trade group which represents cigarette manufacturers.

## Tobacco Products Liability Project
Northeastern University School of Law
400 Huntington Avenue
Boston, MA 02115
(617) 437-2026
*Richard Daynard, Chairperson*

Encourages lawsuits against the tobacco industry to compensate victims of tobacco-related diseases.

## Transportation Alternatives
P.O. Box 2087
New York, NY 10009
(212) 334-9343
*Leona Gonsalves, Administrative Director*

Bicycle commuters, bike messengers, and environmentalists united to support responsible alternatives to the automobile.

## United Citizens Coastal Protection League
P.O. Box 46
Cardiff-by-the-Sea, CA 92007
(619) 753-7477
*Robert Bonde, Executive Director*

Protested the California Bullet Train, a proposed high-speed rail service on the coast. Currently promoting alternative commuter transportation.

**Volkswagen of America**
888 West Big Beaver
Tory, MI 48007
(313) 362-6000
*Hans-Joerg Hungerland, CEO*

**Volvo North American Corporation**
Rockleigh Industrial Park
Rockleigh, NJ 07647
(201) 768-7300
*Bjorn Ahlstrom, CEO*

**World Meteorological Organization**
Case Postale 5
CH-1211 Geneva 20, Switzerland
34 64 00
34 23 26 FAX

Agency of the United Nations created to promote the establishment of a worldwide meteorological observation system for charts and forecasting the earth's climate.

# CHEMICALS

**Problem:** Man-made chemicals pollute our air, land, and water, endangering human, animal, and plant life.

**Solution:** Use natural, non-toxic substances at home and in the workplace. Strictly regulate the use of essential chemicals by industry.

This chapter includes government and non-government groups, international agencies, organizations, companies, and individuals mainly involved with: asbestos, chemical production, fluoride, natural home and personal-care products, organic foods, pesticides, and toxic substance safety. (For other agriculture topics, see *Land.* For disposal issues, see *Recycling and Hazardous Waste.*)

## AFM Enterprises
1140 Stacy Court
Riverside, CA 92507
(714) 781-6860

Manufactures a complete line of products for the environmentally aware, from shampoos to low-toxic, low-odor paints.

## Alexandra Avery
68183 Northrup Creek Road
Birkenfeld/Clatskanie, OR 97016
(503) 755-2446

Company that makes natural, environmentally safe personal care products.

## Allens Naturally
P.O. Box 339
Farmington, MI 48332-0339
(313) 453-5410

Makes a variety of cruelty-free (not tested using animal subjects), non-toxic, bio-degradable home care products.

## American Chemical Society
1155 16th Street, NW
Washington, DC 20036
(202) 872-4600
*Joseph A. Dixon, Chairman of the Board*
*Paul G. Gassman, President*

Society of chemists and chemical engineers.

## American Industrial Hygiene Council
475 Wolf Ledges Parkway
Akron, OH 44311
(216) 762-7294
*Gerald Devitt, Acting Managing Director*

Promotes cleanliness as a means toward safety in the workplace, especially in industries that deal with toxic substances.

**American Institute of Chemical Engineers**
345 East 47th Street
New York, NY 10017
(212) 705-7338
*Dr. Richard E. Emmert, Executive Director*

Largest professional organization of chemical engineers.

**Asbestos Victims of America**
P.O. Box 559
Capitola, CA 95010
(408) 476-3646
*Heather R. Bechtel-Maurer, CEO*

Assists asbestos victims and their families with medical, emotional, and financial problems.

**Aubrey Organics**
4419 Manhattan Avenue
Tampa, FL 33614
(813) 876-4879

100% natural hair and skin care products.

**Aura Cacia**
P.O. Box 399
Weaverville, CA 96093
(916) 623-4999

Manufactures natural, cruelty-free beauty products.

**Autumn-Harp**
28 Rockydale Road
Bristol, VT 05443
(802) 453-4807

Industry leader in people-and-planet conscious business practices; produces personal care products.

**Bio-Integral Resource Center**
P.O. Box 7414
Berkeley, CA 94707
(415) 524-2567
*Sheila Daar, Executive Director*

Established 1979, advocates least toxic pest control. Publishes *The Common Sense Pest Control Quarterly.*

**Body Shop, The**
45 Horsehill Road
Cedar Knolls, NJ 07927-2003
(800) 541-2535
*Anita Roddick, Founder*

Sells all-natural cosmetics.

**Cause for Concern**
159 Belvidere Avenue
Washington, NJ 07882
(201) 689-1392
*Adele T. McIntosh, Spokesperson*

Environmental education organization of parents and consumers concerned with toxic household products.

**Celanese Corporation**
1211 Avenue of the Americas
New York, NY 10036
(212) 719-8000
*John D. Macomber, Chairman of the Board*

Produces petrochemicals, fibers, and plastics.

**Center for Ecological Eating Education**
1377 K Street, NW, Suite 629
Washington, DC 20005
(202) 483-2616

Advocates organically grown food.

**Chemical Referral Center**
c/o Chemical Manufacturers Association
2501 M Street, NW
Washington, DC 20037
(800) 262-8200
*Frances A. Griffin, Manager*

The public interest program of the Chemical Manufacturers Association which serves as a comprehensive chemical information and referral service.

## Citizen's Clearinghouse for Hazardous Waste

P.O. Box 926
Arlington, VA 22216
(703) 276-7070
*Lois Gibbs, Founder*
*Will Collette, Director*

An information service on how to deal with toxic chemicals in your environment.

## Citizens for a Better Environment

33 East Congress, Suite 523
Chicago, IL 60605
(312) 939-1530
*Ron Stevens, President*

Lobbyists working especially in the area of toxic chemical pollutants.

California Office
942 Market Street, #505
San Francisco, CA 94102
(415) 788-0690
*Kevin Shea, Board Chair*

## Concerned Neighbors in Action

P.O. Box 3847
Riverside, CA 92519
(714) 782-4267
*Penny Newman, Chairman*

Conducts advocacy activities for the cleanup of the hazardous waste dump known as the Stringfellow acid pits. Disseminates information nationwide.

## Coordinating Committee on Toxics and Drugs

825 West End Avenue, Suite 70
New York, NY 10025
(212) 663-6378
*Eileen Nic, Program Director*

Exchanges information on hazardous chemical products that are traded internationally.

## Cure Formaldehyde Poisoning Association

Waconia, MN 55387
(612) 442-4665
*Connie Smrecek, Executive Officer*

Seeks to educate legal and health officials concerning problems caused by formaldehyde.

## Dow Chemical Company

2030 Willard H. Dow Center
Midland, MI 48640
(517) 636-1000
*P.R. Oreffice, President*

## Du Pont de Nemours and Company

1007 Market Street
Wilmington, DE 19898
(302) 774-1000
*Edgar S. Woolard, Jr., CEO*

## Earth Tools

9754 Johanna Place
Shadow Hills, CA 91040
(800) 321-9449
(818) 353-5883

Catalog of non-toxic garden products.

## Ecco Bella

6 Provost Square, Suite 602
Caldwell, NJ 07006
(800) 888-5320
*Sally Malanga, President*

Makes bio-degradable cellulose bags (to replace plastic) and personal-care products.

## European Committee for the Protection of the Population Against the Hazards of Chronic Toxicity

c/o Prof. René Truhaut
Laboratoire de Toxicologie, Faculté des Sciences Pharmaceutiques
Université René Descartes
4 avenue de l'Observatoire
F-75006 Paris, France
43 26 71 22

**Everett, Larry**
c/o Banner Press
418 Hudson Street
Hoboken, NJ 07030
(210) 659-4945
Author of *Behind the Poison Cloud,* about the Bhopal tragedy.

**Farm Labor Organizing Committee**
507 South Claire Street
Toledo, OH 43602
(419) 243-3456
*Baldemar Valasquez, President*

A migrant farm workers' union that advocates pesticide control.

**Feather River Company**
133 Copeland
Petaluma, CA 94952
(707) 778-7627

Natural personal care products with bio-degradable packaging.

**Grassroots Environmental Organization**
Box 2018
Bloomfield, NJ 07003
(201) 429-8965

An organization which provides information on how to deal with toxic chemicals in your environment.

**Green Ban**
P.O. Box 745
Longview, WA 98632

Sells natural and safe line of insect repellents and plant insecticides.

**Household Hazardous Waste Wheel**
Box 70
Durham, NH 03824-0070

Provides information to the consumer on alternatives to commercial chemicals.

**I & M Natural Skin Care**
P.O. Box 691, Station "C"
Toronto, Ontario M6J 3S1 Canada
(416) 367-0679

Handmade-to-order natural hair and skin products.

**Institute for Food and Development Policy**
145 Ninth Street
San Francisco, CA 94103
(415) 864-8555
*Thomas Ambrogi, Executive Director*

Published *Circle of Poison: Pesticides and People in a Hungry World.* Concerned with the international use of pesticides on foods.

**Integrated Pest Management**
Cooperative Extension Service
Stockbridge Hall, University of
  Massachusetts
Amherst, MA 01003
(413) 545-2715

Works with Massachusetts growers on pest control.

**International Chemical Workers Union**
1655 West Market Street
Akron, OH 44313
(216) 867-2444
*Frank D. Martino, President*

Union founded in 1940 includes 370 locals with 65,000 members.

**International Organization for Biological Control of Noxious Animals and Pests**
c/o CSIRO Biological Control
  Unit, 335 avenue Parguel
F-34100 Montpelier, France

Promotes chemical-free pest control.

## International Society for Fluoride Research

P.O. Box 692
Warren, MI 48090
(313) 757-2850
*Edith M. Waldbott, Contact*

Research pertaining to the biological and other effects of fluoride on animals, plants, and human life.

## InterNatural

P.O. Box 680, Shaker Street
S. Sutton, NH 03273
(800) 446-4903

Manufactures natural personal care products.

## Lion & Lamb Cruelty-Free Products

29-28 41st Avenue, Suite 813
Long Island City, NY 11101
(718) 361-5757

Manufactures cruelty-free household products which are bio-degradable.

## Monsanto Company

800 North Lindbergh Boulevard
St. Louis, MO 63166
(314) 694-1000
*Richard J. Mahoney, CEO*

Produces plastics, resins, and man-made fibers.

## Morton Thiokol

110 North Wacker Drive
Chicago, IL 60606
(312) 621-5200
*Charles S. Locke, CEO*

Manufactures and markets specialty chemicals.

## Mothers and Others for Pesticide Limits

A Project of the Natural Resources Defense Council
P.O. Box 96641
Washington, DC 20090
(202) 783-7800
*Meryl Streep, Spokesperson*

Launched the successful campaign against Alar, a pesticide for apples.

## National Agricultural Chemicals Association

1155 15th Street, NW, Suite 900
Washington, DC 20005
(202) 296-1585
*Lawrence S. Norton,
Secretary-Treasurer*

Group of firms engaged in producing or formulating agricultural chemical products.

## National Association of Scientific Material Managers

Chemistry Department
University of New Orleans
New Orleans, LA 70148
(504) 286-6324
*Cecil M. Wells, Treasurer*

Publishes a quarterly newsletter covering waste disposal and the relationship between exposure to chemicals and cancer.

## National Coalition Against the Misuse of Pesticides

530 7th Street, SE
Washington, DC 20003
(202) 543-5450
*Jay Feldman, National Coordinator*

Information clearinghouse on toxic free pest control.

## National Farm Workers Health Group

P.O. Box 22579
San Francisco, CA 94122
(415) 731-6569

Farm workers concerned about the effect of pesticides on themselves.

## National Pesticide Telecommunication Network

Department of Preventive Medicine
School of Medicine
Fourth and Indiana
Texas Tech University Health Sciences Center
Lubbock, TX 79430
(806) 743-3091
*Anthony Way, Executive Officer*

Information clearinghouse for pesticides, toxicology and symptomatic reviews, health and environmental effects, safety information, and cleanup and disposal.

## National Toxics Campaign

37 Temple Place, 4th Floor
Boston, MA 02111
(617) 482-1477
*John O'Connor, Coordinator*

Works with victims of toxic poisoning and lobbyists for the Superfund and the Stratospheric Ozone Projection Act.

## Natural Organic Farmers' Association

RFP #2
Barre, MA 01005
(413) 247-9264
*Jerry Fix, President*

Advocates farming methods that show respect for soil, water, and air.

## No More Bhopals Network

c/o Simitu Kothari
Lokayan, 13 Alipur Road
New Delhi 11054, India

Works for continued relief and rehabilitation of the victims of Bhopal.

## Northwest Coalition for Alternatives to Pesticides

P.O. Box 1393
Eugene, OR 97440
(503) 344-5044

A five-state coalition providing an information service.

## Oil, Chemical and Atomic Workers International Union

P.O. Box 2812
Denver, CO 80201
(303) 987-2229
*Joseph Misbrener, President*

Founded in 1955, union of 475 locals with 120,000 members.

## Organic Crop Improvement Association

3185 Township Road, #179
Bellefontaine, OH 43311
(513) 592-4983
*Betty Kananen, Administrative Director*

Seeks to improve organically grown crops through the establishment of an internationally recognized certification process.

## Organic Farms

10726B Tucker Street
Beltsville, MD 20705
(800) 222-6244

One of the largest distributors of organically grown food and also publishes a list of organic restaurants.

## Organic Foods Production Association of North America

226 East Second Street
Winona, MN 55987
(507) 452-6332
*Kate Clancy, Spokesperson*

Helps write laws on organic food production.

## People Against Chlordane

P.O. Box 107
Jericho, NY 11714
*Patrick Menichino, Chairperson*

Strives to ban chlordane, an insecticide believed to cause liver and nervous system damage.

## Pesticide Action Network
P.O. Box 610
San Francisco, CA 94101
(415) 541-9140
*Monica Moore, Executive Director*
*Doria Mueller-Beilschmidt,*
*Information Coordinator*

Opposes the overuse and misuse of pesticides and provides information regarding pesticide use to consumers.

## Public Voice for Food and Health Policy
1001 Connecticut Avenue, NW,
Suite 522
Washington, DC 20036
(202) 659-5930
*Ellen Haas, Executive Director*

Consumer education, research, and advocacy group that promotes fish inspection and pesticide regulation.

## Rachel Carson Council
8940 Jones Mill Road
Chevy Chase, MD 20815
(301) 652-1877
*Samuel S. Epstein, President*

Pesticide information clearinghouse named after the late author of *Silent Spring,* the book which first revealed the toxic pollution of water sources.

## SCM Corporation
299 Park Avenue
New York, NY 10171
(212) 752-2700
*Paul H. Elicker, Chairman*

Produces chemicals, coatings, and resins.

## Sherwin-Williams Company
101 Prospect Avenue, NW
Cleveland, OH 44115
(216) 566-2000
*John G. Breen, CEO*

World's largest producer of paint.

## Silicon Valley Toxics Coalition
760 N. First Street
San Jose, CA 95112
(408) 287-6707

Keeps high-tech electronics firms in line by monitoring groundwater, waste storage, and toxic emissions.

## Society for Agricultural Training Through Integrated Voluntary Activities
Route 2, Box 242W
Viola, WI 54664
(608) 625-2217
*Steven Freer, Director*

Publishes a quarterly newsletter and a listing of organic farms in the Midwest.

## Society for Environmental Geochemistry and Health
Center for Environmental Sciences,
Campus Box 136
University of Colorado
Denver, CO 80204
(303) 556-3460
*Willard R. Chappell, Executive Officer*

Studies the effect of geochemistry on the health of man, animals, and plant life.

## Tom's of Maine
P.O. Box 710, Railroad Avenue
Kennebunk, ME 04043
(207) 985-2944

Natural, bio-degradable personal care products available in many mainstream supermarkets.

## Union Carbide Corporation
Old Ridgebury Road
Danbury, CT 06817
(203) 794-2000
*Warren M. Anderson, CEO*

Produces chemicals and carbon products.

**Universal Proutist Farmers Federation**
1354 Montague, NW
Washington, DC 20011
(202) 882-8804
*Ac. Brahmananda Avt., Chief Secretary*

An organization of small organic farmers. Publishes *Farming the Future.*

**Vietnam Veterans Agent Orange Victims**
205 Main Street
Stamford, CT 06901
(203) 323-7478
*James Sparrow, Executive Director*

Provides medical, legal, and referral services to victims of dioxin poisoning.

**Waltham Field Station**
240 Beaver Street
Waltham, MA 02154
(617) 891-0650

Publishes *Integrated Pest Management Manual for Turf.*

**Weleda**
841 South Main Street,
P.O. Box 769
Spring Valley, NY 10977
(914) 352-6145

Organically grown natural personal care products, including perfume.

**Whelan, Elizabeth, Dr.**
American Council on Science and Health
1995 Broadway, 16th Floor
New York, NY 10023
(212) 362-7044

Chemical industry spokesperson.

**White Lung Association**
1114 Cathedral
Baltimore, MD 21201
(301) 727-6029
*James Site, Executive Director*

Serves as an information clearinghouse for asbestos victims.

**Wrath of Grapes Boycott Information and Support**
United Farm Workers of America, AFL-CIO
La Paz
Keene, CA 93570
(805) 822-5571
*Cesar Chavez, President*

Boycott of grapes organized to focus attention on the use of pesticides.

# EDUCATION

This chapter includes groups which provide educational opportunities on environmental issues, colleges which offer environmental degrees, professional societies, research sources and information clearinghouses, environmental law groups, volunteer organizations, and special programs which teach children and adults.

**ACTION**
1100 Vermont Avenue, NW
Washington, DC 20525
*Donna M. Alvarado, Director*

Principal organizational umbrella for the Peace Corps and VISTA, training volunteers in developmental aid in the U.S. and foreign countries.

Eastern Region
T.P. O'Neill Jr. Federal Bldg.,
10 Causeway
Boston, MA 02202

Midwestern Region
10 West Jackson Boulevard
Chicago, IL 60606

Western Region
11000 Wilshire Boulevard
Los Angeles, CA 90024
(213) 209-7421

**Alaska, University of**
Fairbanks, AK 99775-0990
(907) 474-7600
*David R. Klein, Cooperative
    Wildlife Research Unit*
*James B. Reynolds, Cooperative
    Fishery Research Unit*

Confers degrees in wildlife and fishery research, agriculture and land resource management, etc.

**Alberta, University of**
Occupational Health Program
13-1-3 Clinical Sciences
Edmonton, Alberta T6G 2G3,
    Canada
(403) 432-6403
*Tee L. Guidotti*

Studies effects of the environment on health.

**Alliance for Environmental Education**
2111 Wilson Boulevard, Suite 751
Arlington, VA 22201
(703) 875-8660
*Steven C. Kussman, President*
*Becky Wilson, Secretary*

Advocate for a quality environment through education and training.

**American Academy of Environmental Engineers**
132 Holiday Court, Suite 206
Annapolis, MD 21401
(301) 266-3311
*William C. Anderson, Executive
    Director*

**American Alliance for Health, Physical Education and Dance**
1900 Association Drive
Reston, VA 22091
(703) 476-3400
*Joel Meier, President*
*Doris Corbett, President Elect*

Environmental education pertaining to outdoor recreation.

**American Association of Occupational Health Nurses**
50 Lenox Pointe
Atlanta, GA 30324
(404) 262-1162
*Matilda A. Babbitz, Executive Director*

Nurses who often must confront environmentally caused disease found in the workplace.

**American Institute of Biomedical Climatology**
1023 Welsh Road
Philadelphia, PA 19115
(215) 673-8368
*Richmond G. Kent, Executive Director*

Meteorologists, biologists, physicians, physicists, engineers, architects, and other professionals investigating the influence of the natural environment on man's health.

**American Medical Association (AMA)**
535 North Dearborn Street
Chicago, IL 60610
(312) 464-5000
*James H. Sammons, Executive Vice-President*

Founded in 1847 as the national association for U.S. physicians.

**American Museum of Natural History**
Central Park West at 79th Street
New York, NY 10024
(212) 769-5000
*George D. Langdon, Jr., President*
*William J. Moynihan, Director*

Founded in 1869. Studies ecological relationships.

**American Nature Study Society**
5881 Cold Brook Road
Homer, NY 13077
(607) 749-3655
*Paul Spector, President*

Environmental education and advocacy.

**American Planning Association**
1776 Massachusetts Avenue, NW
Washington, DC 20036
(202) 872-0611
*Israel Stollman, Executive Director*

Concerned with the planned development of urban and rural communities.

**American Public Health Association**
1015 15th Street, NW
Washington, DC 20005
(202) 789-5600
*William H. McBeath, Executive Director*

**American Public Works Association**
1313 East 60th Street
Chicago, IL 60637
(312) 667-2200
*Robert D. Bugher, Executive Director*

**American Society for Environmental Education**
1200 Clay Street, #2
San Francisco, CA 94108
(415) 474-7123
*Alan T. Wedwick, Contact*

Purpose is to foster environmental education in schools and with the public.

## American Society for Environmental History
Center for Technology Studies
New Jersey Institute of Technology
Newark, NJ 07102
(201) 596-3270
*William Cronon, President*

Publishes *Environmental Review* and specializes in human ecology.

## American Society of Civil Engineers
345 East 45th Street
New York, NY 10017
(212) 705-7496
*Edward O. Pfrang, Executive Director*

## Arctic Institute of North America
University Library Tower
2500 University Drive, NW
Calgary, Alberta T2N 1N4 Canada
(403) 220-7515
*Michael Robinson, Executive Director*

Extensive information, student research on Arctic areas.

## Arizona State University
Tempe, AZ 85287

Offers degrees in environmental subjects.

## Association of Environmental Engineering Professors
Department of Civil Engineering
University of California, Davis
Davis, CA 95616
(916) 752-1440
*George Tchobanogolus, President*

## Augustana Research Institute
Box 783, Augustana College
Sioux Falls, SD 57056
(605) 336-4912
*John Sorenson, Chairman*

Conducts investigations of a pure and applied nature in the northern Great Plains region.

## Bolton Institute for a Sustainable Future
4 Linden Square
Wellesley, MA 02181-4709
(617) 235-5320
*Elizabeth and David Dodson Gray, Co-Directors*

Educates the public on growth policy.

## Boston College Environmental Affairs Law Review
Boston College Law School
885 Centre Street
Newton Centre, MA 02159
(617) 552-4350

## Boy Scouts of America
P.O. Box 152079, 1325 Walnut Hill Lane
Irving, TX 75015-2079
(214) 580-2000
*Harold S. Hook, President*
*Ben H. Love, Chief Scout Executive*

Established in 1910, teaches scouts to treat the environment with respect and care: Always leave a place better than you found it.

East Central Region
230 West Dieh Road
Naperville, IL 60540
(312) 983-6730
*M. Gene Cruse, Regional Director*

North Central Region
P.O. Box 29100
Overland Park, KS 66201
*J. Thomas Ford, Jr., Regional Director*

Northeast Region
P.O. Box 350
Dayton, NJ 08810
(201) 821-6500
*Rudolph Flythe, Regional Director*

South Central Region
P.O. Box 15235
Irving, TX 75015
(214) 580-2471
*Richard O. Bentley, Regional Director*

Southeast Region
P.O. Box 440728
Kennesaw, GA 30144
(404) 955-2333
*C. Hoyt Hunt, Regional Director*

Western Region
P.O. Box 3464
Sunnyvale, CA 94088-3464
(408) 735-1201
*Richard R. Harrington, Regional Director*

## California, University of

Berkeley
Forestry and Resources
Management
Berkeley, CA 94720
(415) 642-0376
*John A. Helms, Chairman*

Offers undergraduate and graduate degrees.

Berkeley
Environmental Science
Department
Berkeley, CA 94720
(415) 642-2628
Interdisciplinary undergraduate program.

Davis
Environmental Studies
Davis, CA 95616
(916) 752-6586
*C. Goldman, Chairman*
Offers graduate and
undergraduate degrees.

Los Angeles
Analysis and Conservation of
Ecosystems
Los Angeles, CA 90024
(213) 825-4321
*Hartmut Walter, Chairman*

Riverside
Environmental Science
Riverside, CA 92521
(714) 787-4551
*John Letey, Jr., Chairman*
Offers undergraduate degrees.

Santa Cruz
Environmental Studies
Santa Cruz, CA 95064
(408) 429-2634
*Michael Soule, Chairperson*
Confers an environmental studies
degree.

## California Conservation Corps

1530 Capitol Avenue
Sacramento, CA 95814
(916) 445-0307
*Bud Sheble, Director*

Corps has a dual mission: conservation
and the employment of the state's
youth.

## California Department of Education

Conservation Education Service
721 Capitol Mall, P.O. Box 944272
Sacramento, CA 94244-2720
(916) 324-7190
*Thomas P. Sachse, Education Unit Manager*

## California Institute of Public Affairs

P.O. Box 10
Claremont, CA 91711
(714) 624-5212

Affiliated with Claremont Colleges,
conducts research and policy forums
and publishes reference books on environmental and natural resources issues.

## Camp Fire
4601 Madison Avenue
Kansas City, MO 64112
(816) 756-1950
*Tish Spaulding Oden, National
President*

Boys and Girls Club, founded 1910. Teaches children, from birth to age 21, about care, protection, and usage of outdoors.

## Canadian Environmental Law Association
243 Queen Street West, Fourth Floor
Toronto, Ontario M5V 1Z4
Canada
(416) 977-2410

## CEIP Fund
68 Harrison Avenue
Boston, MA 02111
(617) 426-4375
*John R. Cook, Jr., President*

Publishes *The Complete Guide to Environmental Careers.* Formerly the Center for Environmental Intern Programs.

## Center for Environment, Commerce and Energy
2733 6th Street, SE, Suite 1
Washington, DC 20003
(202) 543-3939
*Norris McDonald, President*

Provides opportunities for blacks and other minorities to participate in environmental activities.

## Center for Environmental Information
99 Court Street
Rochester, NY 14604
(716) 546-3796
*Elizabeth Thorndike, President*

Information resource with an extensive library.

## Center for Environmental Study
Grand Rapids Junior College
143 Bostwick, NE
Grand Rapids, MI 49503
(616) 456-4848
*Kay T. Dodge, Executive Director*

Provides consulting services for international education and research.

## Center for Holistic Resource Management
P.O. Box 7128
Albuquerque, NM 87194
(505) 242-9272
*Stan Kano, Executive Director*

Focuses on holistic management of land, water, human, wildlife, and financial resources.

## Center for International Development and Environment
1709 New York Avenue, NW
Washington, DC 20006
(202) 462-0900
*Thomas Fox, Director*

Seeks to achieve balance between the long-term conservation of natural resources, the environment, and human needs.

## Center for Science Information
4252 20th Street
San Francisco, CA 94114
(415) 553-8772
*Steven C. Witt, President*

Educates journalists and decision makers on the environmental and public policy issues surrounding biotechnology.

## Children for Old Growth
Box 1090
Redway, CA 95560
(707) 923-3617

Focuses on encouraging children to help save the remaining virgin forests.

**Colorado School of Mines**
Department of Environmental
Sciences
Golden, CO 80401


**Colorado University**
**Environmental Center**
Fort Collins, CO 80523


**Commission on Professionals in**
**Science and Technology**
1500 Massachusetts Avenue, NW,
Suite 831
Washington, DC 20005
(202) 223-6995
*Betty M. Vetter, Executive Director*

Seeks to aid the development of U.S.
scientific resources and promotes scientific training.


**Conservation and Research**
**Foundation**
Connecticut College
New London, CT 06320
(203) 873-8514
*Richard H. Goodwin, President*

Encourages study and research in the
biological sciences and promotes conservation of natural resources.


**Conservation Districts**
**Foundation**
Conservation Film Service, Davis
Conservation Library
404 E. Main, P.O. Box 855
League City, TX 77574-0855
(713) 332-3404
*Ernest Shea, Executive*
*Vice-President*

Collects, catalogs, and loans materials
regarding the conservation movement
in America.


**Conservation Education**
**Association**
c/o Oklahoma Conservation
Commission
2800 Lincoln, Suite 160
Oklahoma City, OK 73105
(405) 521-2384
*Dan Sebert, Executive*
*Secretary-Treasurer*

Disseminates news, ideas, and suggestions for local, state, and national conservation education programs.


**Conservation Law Foundation**
**(CLF)**
3 Joy Street
Boston, MA 02108
(617) 742-2540
*Francis W. Hatch, Jr., Chairman of*
*the Board*


**Consumers Union of United**
**States**
256 Washington Street
Mt. Vernon, NY 10553
(914) 667-9400
*Rhoda H. Karpatkin, Executive*
*Director*

Publishes *Consumer Reports* and tests,
rates, and provides information on
household goods and other products.
Founded in 1936.


**Coolidge Center for**
**Environmental Leadership**
1675 Massachusetts Avenue,
Suite 4
Cambridge, MA 02138-1836
(617) 864-5085
*Peter J. Ames, Chairman of the*
*Board*

Programs for future leaders of developing countries.


**Dag Hammarskjold Foundation**
Ovre Slottsgatan 2
S-752 20 Uppsala, Sweden
10 54 72

Organizes invitational seminars and conferences on the world environment.

## Earth Ecology Foundation
2305 East Ashlan Avenue
Fresno, CA 93726
(209) 224-0478
*Erik Wunstell, Director*

Purpose is to study the relationship among humans, technology, and nature.

## Earth Society Foundation
585 Fifth Avenue
New York, NY 10017
(718) 574-3059
*John McConnell, Chairman*

Acts as an information clearinghouse and supports groups concerned with environmental care and protection of the earth.

## Earthmind
P.O. Box 743
Mariposa, CA 95338
*Michael A. Hackleman, President*

Research and education concerning alternative energy sources and organic gardening.

## Earthwatch
680 Mt. Auburn Street
Watertown, MA 02172
(617) 926-8200
*Brian A. Rosborough, President*

An amateur group of researchers led by professional scientists whose studies include ecology.

## Ecological Society of America
c/o Department of Botany
University of Wyoming
Laramie, WY 82071
(307) 766-3291
*Dennis H. Knight, President*

Society of ecologists providing perspectives on areas of public policy.

## Eleventh Commandment Fellowship
P.O. Box 14667
San Francisco, CA 94114
(415) 626-6064
*Frederick W. Krueger, Coordinator*

Christian ecologists urging an 11th Commandment: "The earth is the Lord's and the fullness thereof; thou shall not despoil the earth nor destroy the life thereon."

## Elmwood Institute
P.O. Box 5805
Berkeley, CA 94705
(415) 845-4595
*Frank Seal, Executive Director*

Promotes "Ecothinking," a philosophy of global interdependence and ecological wisdom.

## Emory University
Atlanta, GA 30322
(404) 329-6292

Offers graduate degrees in terrestrial ecology and systems ecology.

## Energy, Environment, and Resources Center
University of Tennessee
327 South Stadium Hall
Knoxville, TN 37996-0710
(615) 974-4251
*Dr. E. William Colglazier, Director*

Does research on environmental and energy programs, serving as a problem solver for the government and the private sector.

## Environmental Action Foundation
1525 New Hampshire Avenue
Washington, DC 20036
(202) 745-4870
*Ruth Caplan, Director*

Public foundation formed to develop research and conduct educational programs.

**Environmental Education Coalition**
Pocono Environmental Education Center
R.D. 2, Box 1010
Dingmans Ferry, PA 18328
(717) 828-2319
*John J. Padalino, Co-Chairman*
*John Paulk, Co-Chairman*

Attempts to upgrade environmental education through increasing citizen literacy and improving environmental experiences for students.

**Environmental Industry Council**
1825 K Street, NW, Suite 210
Washington, DC 20006
(202) 331-7706
*Milt Capps, Executive Director*

Association of manufacturers of pollution control equipment.

**Environmental Law Institute**
1616 P Street, NW, Suite 200
Washington, DC 20036
(202) 328-5150
*J. William Futrell, President*

Educates on environmental law and policy.

**Envirosouth, Inc.**
P.O. Box 11468
Montgomery, AL 36111
(205) 277-7050
*Martha McInnis, President*

Public information service on environmental concerns in Alabama.

**ERIC Clearinghouse for Science, Mathematics, and Environment Education—Ohio State University**
School of Natural Resources
1200 Chambers Road (3rd Floor)
Columbus, OH 43212
(614) 292-2265

Database of environmental education resources.

**Evergreen State College**
Olympia, WA 98505
(206) 866-6000

Offers environmental studies and training programs at the undergraduate and graduate levels.

**Family Games, Inc.**
P.O. Box 97
Snowdon, Montreal, Quebec H3X 3T3 Canada
(514) 485-1834

Produces a line of ecological board games for adults and children.

**Futurepast: The History Company**
North 10 Post Street, Suite 550,
P.O. Box 1905
Spokane, WA 99210-1905
(509) 836-5242
*John C. Shideler, President*

Environmental history experts.

**George Washington University**
Washington, DC 20052
(202) 994-1000

Offers degrees in environmental law, environmental studies, environmental engineering and environmental management.

**Georgia, University of**
Institute of Ecology
Athens, GA 30602
(404) 542-2968
*Michael J. Van Den Avyle, Leader*
*L. A. Hargreaves, Jr., Dean*

Confers degrees in wildlife, fisheries, natural resources, etc.

**Georgia Institute of Technology**
Environmental Resources Center
Atlanta, GA 30332
(404) 894-3776
*Bernd Kahn, Director*

## Georgia Marine Institute, University of

Sapelo Island, GA 31327
(912) 485-2221
*James J. Alberts, Director*

Concerned with research into the system-ecology, biology, chemistry, and geology of salt marshes.

## Girl Scouts of the United States of America

830 3rd Avenue
New York, NY 10022
(212) 940-7500
*Betty Pillsbury, President*

Education in conservation, wildlife protection, community service.

## Global Exchange

2940 16th Street, #307
San Francisco, CA 94103
(415) 255-7296

Travel company that encourages ecotourism—socially conscious touristing.

## Goede Educational Options

P.O. Box 106, 301 South Church Street
West Chester, PA 19381
(215) 692-0413

Publishes environmental books and games for children.

## Green Library

1918 Bonita Avenue
Berkeley, CA 94704
(415) 845-9975

Works to establish environmental libraries.

## Group for Environmental Education

1214 Arch Street
Philadelphia, PA 19107
(215) 564-4403

Promotes environmental education on primary, secondary, and advanced levels.

## Harvard Environmental Law Review

Harvard Law School
Cambridge, MA 02138
(617) 495-3110

## Harvard University

212B Pierce Hall
Cambridge, MA 02138
(617) 495-2833

Offers a degree in engineering sciences with an environmental sciences emphasis as well as a one-year program in forest science.

## Hawaii Evolutionary Biology Program

University of Hawaii/Manoa
3050 Maile Way
Honolulu, HI 96822
(808) 956-6739

## Hudsonia Limited

Bard College
Annandale, NY 12504
(914) 758-1881
*Erik Kiviat, Executive Director*

Research, education, and consulting in environmental sciences for the Hudson River region.

## Human Environment Center

1001 Connecticut Avenue, NW, Suite 827
Washington, DC 20036
(202) 331-8387
*David Burwell, Co-Chair of the Board*
*Shirley Malcom, Co-Chair*

Clearinghouse and technical assistance center for youth conservation and service corps programs.

## Humboldt State University

Arcata, CA 95521
(707) 826-3561
*Richard L. Reidenhour, Dean*

The College of Natural Resources offers degrees in fisheries, forestry, oceanography, etc.

**Idaho, University of**
College of Forestry, Wildlife, and
Range Sciences
Moscow, ID 83843
(208) 236-3765
*Rod R. Seeley, Chairman*

Offers undergraduate and graduate degrees.

Range Resources
Twin Falls, ID 83301
(208) 885-6536
*Kendall Johnson, Director*
Offers undergraduate and
graduate degrees.

**Idaho State University**
Department of Biological Sciences
Box 8007
Pocatello, ID 83209
(208) 885-6441
*Rod R. Seeley, Chairman*

Grants undergraduate and graduate
degrees in ecology.

**Illinois, University of, at
Urbana-Champaign**
Institute for Environmental Studies
Urbana, IL 61801
(217) 333-1000
*Roger A. Minear, Director*

Offers undergraduate and graduate degrees.

**Indiana University**
School of Public and
Environmental Affairs
Terre Haute, IN 47809
(812) 335-9485
*Charles F. Bonser, Dean*

Offers undergraduate and graduate degrees in environmental affairs.

**Inform, Inc.**
381 Park Avenue, South
New York, NY 10016
(212) 689-4040
*Kenneth F. Mountcastle, Jr.,
Chairman of the Board*

Education and research identifying and
reporting on practical actions for conservation.

**Institute for Conservation
Leadership**
1400 16th Street, NW
Washington, DC 20036-2266
(202) 797-6656
*Ed Easton, Director*

Training for volunteer conservation
leaders.

**Institute for Earth Education**
Box 288
Warrenville, IL 60555
(312) 393-3096
*Steve Van Matre, Chair*

International institute of environmental educators, naturalists, teachers,
youth workers, camp counselors, etc.

**International Association for
Ecology**
c/o Institute of Ecology
University of Georgia
Athens, Georgia 30602
(404) 542-2968
*Frank B. Golley, President*

The international organization of professional ecological scientists.

**International Association for
the Advancement of Earth
and Environmental Sciences**
Geography & Environmental
Studies Department
5500 North St. Louis Avenue
Chicago, IL 60625
(312) 794-2628
*Musa Qutub, President*

Seeks to improve and encourage the
study of science at elementary, secondary, and university levels.

**International Council for
Outdoor Education**
P.O. Box 17255
Pittsburgh, PA 15235
(412) 372-5992
*John F. Eveland, President*

## International Council of Environmental Law
Adenaueralle 214
D 53 Bonn, Germany
49-228-269240
*W. E. Burhenne, Executive Governor*

## Iowa State University
Department of Animal Ecology
124 Science Building II
Ames, IA 50011
(515) 294-4111

## John Muir Institute for Environmental Studies
743 Wilson Street
Napa, CA 954559
(707) 252-8333
*Max Linn, President*

Identifies and studies environmental problems that are not receiving large-scale attention from other organizations.

## Johns Hopkins University
Baltimore, MD 21218
(301) 338-8000

Offers degrees in environmental engineering, environmental health sciences, etc.

## Kansas, University of
Environmental Studies
Lawrence, KS 66045
(913) 864-4301
*Frank Denoyelles, Director*

Offers undergraduate and graduate degrees.

## Keystone Center
P.O. Box 606
Keystone, CO 80435
(303) 468-5822
*Paul A. Downey, Chairman*

Conducts national policy dialogues on the environment.

## Lewis and Clark College
Northwestern School of Law
10015 SW Terwilliger Boulevard
Portland, OR 97219
(503) 768-6600

Publishes *Environmental Law Journal.*

## Manitoba Naturalists Society
302-128 James Avenue
Winnipeg, Manitoba R3B ON8
Canada
(204) 943-9029
*Rod Tester, President*

Founded in 1920 to foster an awareness and appreciation of the natural environment.

## Mark Trail/Ed Dodd Foundation
P.O. Box 2907
Gainesville, GA 30503
(404) 532-4274
*Rosemary Dodd, Executive Director*

Teaches children conservation through comic strip character Mark Trail.

## Maryland, University of
Center for Environmental and Estuarine Studies
Cambridge, MD 21613
(301) 228-9250
*Thomas C. Malone, Director*

## Massachusetts Institute of Technology
Room 3-234
Cambridge, MA 02139
(617) 253-7753
*Louis Menand, Office of the Provost*

## Miami, University of
Rosenstiel School of Marine and Atmospheric Science
4600 Rickenbacker Causeway
Miami, FL 33149
(305) 361-4000
*Christopher G. A. Harrison, Dean*

## Michigan State University
East Lansing, MI 48824
(517) 355-1855

Offers degrees in forestry, fisheries/wildlife and resource development.

## Minnesota, University of
Hubert H. Humphrey Institute of Public Affairs
Minneapolis, MN 55155
(612) 373-2653
*Dean E. Abrahamson, Director of the Global Environmental Policy Project*

## Missouri Department of Conservation
Association for Conservation Information
408 South Polk
Albany, MO 64402
(816) 726-3677
*Rod Green, President*

Assists in setting up conservation programs and facilitates free exchange of materials.

## Music for Little People
1144 Redway Drive
Redway, CA 95560
(707) 923-3991

A catalog of environmentally conscious tapes, videotapes, and storybooks for children.

## National Association of Biology Teachers
11250 Roger Bacon Drive, #19
Reston, VA 22090
(703) 471-1134
*Nancy Ridenour, President*

## National Association of Environmental Professionals
P.O. Box 15210
Alexandria, VA 22309
(703) 660-2364
*Joan A. Schroeder, Executive Secretary*

Goal is to improve communications and advance the environmental planning process.

## National Association of Interpretation
P.O. Box 1892
Fort Collins, CO 80522
(303) 491-6434
*Paul Frandsen, President*

Advances skills of those who act as guides in parks and museums.

## National Association of Service and Conservation Corps
1001 Connecticut Avenue, Suite 827
Washington, DC 20036
(202) 331-9647
*Robert Burkhardt, Board President*

## National Council for Environmental Balance
4169 Westport Road, P.O. Box 7732
Louisville, KY 40207
(502) 896-8731
*I. W. Tucker, President*

Dedicated to a balanced approach to solving environmental and energy problems without destroying the economy "and people's rights to a responsible life."

## National Education Association (NEA)
1201 16th Street, NW
Washington, DC 20036
(202) 833-4000
*Keith Geiger, President*

Promotes environmental education.

## National Science Teachers Association
1742 Connecticut Avenue, NW
Washington, DC 20009
(202) 328-5800
*LaMoine L. Motz, President*

Hopes to improve the teaching of science, pre-school through college.

## Natural Science for Youth Foundation
130 Azalea Drive
Roswell, GA 30075
(404) 594-9367
*John Ripley Forbes, Chairman of the Board*

Provides information services and educational tools.

## New England Field Guide to Environmental Education Facilities and Resources
New England Field Guide,
Antioch/NE Graduate School
103 Roxbury Street
Keene, NH 03431
(603) 357-3122

## New England Natural Resources Center
200 Lincoln Street
Boston, MA 02111
(617) 451-3670
*Perry Hagenstein, Chairman*

A trust organized to provide a focal point for discussion and resolution of regional natural resource and environmental issues.

## New Mexico, University of
School of Law, Natural Resources Center
Albuquerque, NM 87131
(505) 277-0111

Includes the International Transboundary Resource Center which focuses on U.S./Mexico problems.

## North American Association for Environmental Education
P.O. Box 400
Troy, OH 45373
(513) 698-6493
*Joan C. Heidelberg, Executive Vice-President*

Promotes environmental education on all levels.

## North Dakota Institute for Ecological Studies
University of North Dakota
P.O. Box 8278, University Station
Grand Forks, ND 58202
(701) 777-2851
*Rodney D. Sayler, Director*

## Planet Drum
Box 31251
San Francisco, CA 94131
(415) 285-6556
*Peter Berg, Founder*

The first group to propose bio-regionalism, an approach to saving the planet by restoring one's own region.

## Project Learning Tree
1250 Connecticut Avenue, NW, Suite 320-FG
Washington, DC 20036
(202) 463-2468
*Lawrence D. Wiseman, President*

Conducts programs that train teachers to teach children about the environment.

## Public Environment Center
One Milligan Place
New York, New York 10011
(212) 691-4877
*H. Alan Hoglund, President*

Broad-ranging information center.

## Purdue University
Department of Forestry and Natural Resources
West Lafayette, IN 47907
(317) 494-3591
*Dennis C. Le Master, Head*

Confers undergraduate and graduate degrees in forestry, wildlife, and fisheries.

## René Dubos Center for Human Environments
100 East 85th Street
New York, NY 10028
(212) 249-7745
*Ruth A. Eblen, Executive Director*

Aids businesses, the public, and government in the environmental decision-making process. Named for scientist/environmentalist René Dubos.

## Renewable Natural Resources Foundation
5430 Grosvenor Lane
Bethesda, MD 20814
(301) 493-9101
*Nancy Hill, Director of Administration and Finance*

A coalition of natural resource groups.

## Resource-Use Education Council
P.O. Box 10026
Richmond, VA 23240
(804) 782-2457
*Helen Jeter, Chairman*

A volunteer organization whose members include state and federal government officials, educators and industrialists, promoting environmental education.

## Rhode Island, University of
Department of Natural Resources Sciences
Kingston, RI 02881
(401) 792-2495
*William R. Wright, Chairman*

Specializes in wetlands and deep-water habitats.

## Scientists' Institute for Public Information
355 Lexington Avenue, 16th Floor
New York, NY 10017
(212) 661-9110
*Alan McGowan, President*

Provides objective scientific data on current social issues linked to science and technology.

## Smithsonian Institution
National Museum of Natural History/National Museum of Man
10th Street and Constitution Avenue, NW
Washington, DC 20560
(202) 357-1300
*Frank Hamilton Talbot, Director*

## Society for Conservation Biology
Biology Department
Montana State University
Bozeman, MT 59717
(406) 994-4548

National society of conservation biologists.

## Society for Educational Reconstruction
c/o Department of Education
University of Bridgeport
Bridgeport, CT 06602
(203) 624-3687
*T. M. Thomas, Director*

A group of university teachers who educate in a way that shows concern for the ecological and the humanitarian.

## Society for Occupational and Environmental Health
P.O. Box 42360
Washington, DC 20015
(202) 797-8666
*Paula Cohen, Executive Director*

Seeks to improve the quality of both working and living conditions by studying hazards and health effects.

## Soil and Water Conservation Society

7515 N.E. Ankeny Road
Ankeny, IA 50021
(515) 289-2331
*Alan C. Epps, Executive Vice-President*

Soil and water conservationists dedicated to the advancement of good land and water use.

## Southern California, University of

Environmental Engineering Program
Los Angeles, CA 90089-0231
(213) 743-7517
*Mihran S. Agbabian, Director*

Confers graduate degrees.

## State University of New York

College of Environment Science and Forestry
123 Bray Hall
Syracuse, NY 13210
(315) 470-6644
*Rod Cochran, Head*

Offers graduate and undergraduate degrees.

## STOP

716 St. Ferdinand Street
Montreal, Quebec H4C 2T2
Canada
(514) 932-7267
*Lynnae Dudley, President*

Broad-based environmental education group devoted to improving the quality of the physical environment.

## Student Conservation Association

Box 550
Charlestown, NH 03603
(603) 826-5206
*Henry S. Francis, Jr., President*

Supporters of the U.S. Forest Service program to enlist students in the preservation of national parks and forests.

## Thorne Ecological Institute

5370 Manhattan Circle, Suite 104
Boulder, CO 80303
(303) 499-3647
*Susan Q. Foster, Executive Director*

Consultants dedicated to improving the human environment through ecological principles via seminars and information publishing.

## Threshold, Inc.

International Center for Environmental Renewal
Drawer CU
Bisbee, AZ 85603
(602) 432-7353
*John Milton, President*

Develops ecologically sound and practical alternatives in conservation.

## Washington, University of

Institute for Environmental Studies
FM-12
Seattle, WA 98195
(206) 543-1812
*Gordon H. Orians, Director*

Confers graduate and undergraduate degrees and sponsors research.

## West Florida, University of

Department of Biology
Pensacola, FL 32514-5751
(904) 474-2746

Offers specialized master's degree in coastal zone studies.

## Western Regional Environmental Education Council

c/o Don Carter, New Mexico Game and Fish
State Capitol Building
Santa Fe, NM 87503
(505) 827-7911
*Peggy Cowan, President*

Association of 13 western states.

**Wisconsin Association for Environment Education**
2428 Downy
Green Bay, WI 54303
(414) 499-6500
*Randell Champeau, Chairperson*

**Wisconsin Department of Public Instruction**
125 S. Webster Street, P.O. Box 7841
Madison, WI 53707
(608) 267-9266
*David C. Engleson, Supervisor of Environmental Education*

Promotes environmental education in public schools.

**Women's Occupational Health Resource Center**
117 St. John's Place
Brooklyn, NY 11217
(718) 230-8822
*Jeanne M. Stellman, Executive Director*

Clearinghouse for women's occupational health and safety issues.

**Woods Hole Oceanographic Institution**
Woods Hole, MA 02543
(508) 548-1400
*Robert Ballard, Senior Scientist*

Includes the world's oldest fisheries research facility, the Sea Education Association, and a National Marine Fisheries Service station.

**World Future Studies Federation**
c/o Social Science Research Institute
University of Hawaii
Honolulu, HI 96822
(808) 948-6601
*James A. Dator, Secretary General*

Holds regional and global future studies conferences.

**World Resources Institute**
1735 New York Avenue, NW
Washington, DC 20006
(202) 638-6300
*James Gustave Speth, President*

Policy research center tackling major questions of environmental protection and industrial growth.

**World Women in the Environment**
1250 24th Street, NW, Suite 500
Washington, DC 20037
(202) 331-9863
*Cynthia R. Helms, President*

Mobilizes women to maintain and improve the environment.

**Yale University**
School of Forestry and Environmental Studies
205 Prospect Street
New Haven, CT 06511
(203) 432-5109

Offers graduate degrees.

**Young Naturalist Foundation**
59 Front Street, E
Toronto, Ontario M5E 1B3
Canada
(416) 868-6001

Publishes *Chickadee*, a wildlife magazine for children.

**Zoo Books**
Box 85271, Suite 6
San Diego, CA 92138
(619) 299-5034

Publishes a series of environmentally oriented books about animals for children.

# ENERGY

**Problem:** The internal combustion engines in our cars, trucks, buses, and planes use petroleum, causing half of all man-made air pollution. We derive much of our energy from fossil fuels, such as natural gas and coal, or nuclear sources. The pollution, waste disposal, and safety concerns associated with all of these forms of energy are serious hazards to the environment.

**Solution:** Lower fuel consumption with increased public transportation, higher mileage efficiency, diminished reliance on fossil fuels, and production of alternative energy sources which are cleaner and more efficient.

This chapter includes government and non-government groups, international agencies, organizations, companies, and individuals mainly involved with: alcohol fuels, coal, economic issues regarding energy, electric utilities, energy efficiency, fusion power, hydropower, natural gas, nuclear, petroleum, solar, and wind energy. (For some of these topics, also see *Air.*)

## Abalone Alliance
2940 16th Street, Room 310
San Francisco, CA 94103
(415) 861-0592

Promotes alternative energy sources, which will decrease nuclear dependency.

## Alliance of Atomic Veterans
Box 32
Topock, AZ 86436
(602) 768-7515
*Anthony Guarisco, Director*

Made up of individuals from U.S., England, Canada, and Australia who were exposed to nuclear weapons explosions.

## Alliance to Save Energy
1725 K Street, NW, Suite 914
Washington, DC 20036
(202) 857-0666
*James L. Wolf, Executive Director*

Coalition of business, government, and consumer leaders who seek to increase the efficiency of energy use.

## Alternative Energy Resources Organization
44 North Last Chance Gulch,
#8/9
Helena, MT 59601
(406) 443-7272
*Al Kurki, Executive Director*

Provides research and public education on sustainable agriculture and renewable energy sources.

## Alternative Sources of Energy
107 South Central Avenue
Milaca, MN 56353
(612) 983-6892
*Donald Marier, Director*

Primarily concerned with wind power, hydropower, and photovoltaics promoted by the independent power production industry.

## American Council for an Energy Efficient Economy
1001 Connecticut Avenue, NW, Suite 535
Washington, DC 20036
(202) 429-8873
*Carl Blumstein, Managing Director*

Encourages the implementation of more energy efficient and economical technologies and practices.

## American Gas Association
1515 Wilson Boulevard
Arlington, VA 22209
(703) 841-8400
*George H. Lawrence, President*

An alliance of natural, manufactured, and liquified gas providers.

## American Institute of Physics
335 East 45th Street
New York, NY 10017-3483
(212) 661-9404
*Kenneth W. Ford, Executive Director*

## American Nuclear Energy Council
410 1st Street, SE
Washington, DC 20003
(202) 484-2670
*Edward M. Davis, President*

Supports nuclear power as an energy source.

## American Nuclear Society
555 North Kensington Avenue
La Grange Park, IL 60525
(312) 352-6611
*Octave J. Du Temple, Executive Director*

Scientists and professionals dedicated to advancing science and engineering in the nuclear industry.

## American Petroleum Institute
1220 L Street, NW
Washington, DC 20005
(202) 682-8000
*Charles J. Dibona, President*
*William F. O'Keefe, CEO*

Cooperates with government and the conservation community for the wise use of energy compatible with the environment.

## American Solar Energy Association
1667 K Street, NW, Suite 395
Washington, DC 20006
(202) 347-2000
*John Lillard, General Counsel*

Distributors, retailers, engineers, and architects in the solar energy field.

## American Solar Energy Society
2400 Central Avenue, B-1
Boulder, CO 80301
(303) 443-3130
*Larry Sherwood, Director*

Professionals organized to promote solar energy through science and technology.

## American Wind Energy Association
1730 North Lynn Street, Suite 610
Arlington, VA 22209
(703) 276-8334
*Randall Swisher, Executive Director*

Promotes wind as an alternative energy source.

## Americans for Energy Independence
1629 K Street, NW, Suite 500
Washington, DC 20006
(202) 466-2105
*Elihu Bergman, Executive Director*

Objectives are to develop and utilize available domestic energy resources as an essential condition for economic health and national security.

## Americans for Nuclear Energy
2525 Wilson Boulevard
Arlington, VA 22201
(703) 528-4430
*Douglas O. Lee, Chairman*

Lobbies state and federal governments in favor of nuclear energy.

## Amoco
200 East Randolph Drive
Chicago, IL 60601
(312) 856-6111
*Richard M. Morrow, CEO*

## Atlantic Richfield Company (ARCO)
515 South Flower Street
Los Angeles, CA 90071
(213) 486-3511
*Lowdrick M. Cook, CEO*

## Biomass Energy Research Association
1825 K Street, NW, Suite 503
Washington, DC 20006
(202) 785-2856
*Donald L. Klass, President*

Facilitates technology transfer, information exchange and education in biomass energy research.

## British Petroleum (BP)
Britannica House, Moor Lane
London EG2Y 9BU, England
*Peter Walter, CEO*

## California Energy Commission
1516 Ninth Street
Sacramento, CA 95814
(916) 324-3000
*Charles R. Imbrecht, Chairman*

State agency that ensures a reliable and affordable supply of energy for California.

## Center for Alternative Mining Development Policy
210 Avon Street, #9
La Crosse, WI 54603
(608) 784-4399
*Al Gedicks, Executive Director*

Seeks to provide information and technical assistance to Indian tribes and rural communities affected by plans for mining development and radioactive waste disposal.

## Center for Energy Policy and Research
c/o New York Institute of Technology
Old Westbury, NY 11568
(516) 686-7578
*Gale Tene Spak, Dean*

Conducts research into energy utilization and conservation.

## Chevron Corporation
225 Bush Street
San Francisco, CA 94104
(415) 894-7700
*George M. Keller, Chairman*

## Citizen/Labor Energy Coalition
225 West Ohio, Suite 250
Chicago, IL 60610
(312) 875-5153

A consumer organization that deals with toxic waste, renewable fuels, health care, and insurance reform.

## Citizens Energy Council
77 Homewood Avenue
Allendale, NJ 07401
(201) 327-3194
*Larry Bogart, National Coordinator*

The oldest opposition group to nuclear power and nuclear weapons, founded in 1966.

## Colorado Office of Energy Conservation

112 East 14th Avenue
Denver, CO 80203
(303) 866-2057
*Judy Harrington, Director*

State agency that promotes conservation and helps citizens to use energy more efficiently.

## Committee for Nuclear Responsibility

P.O. Box 11207
San Francisco, CA 94101
*John W. Gofman, Chairman*

Offers comprehensible, practical, and independently funded analysis of the effects upon cancer and genetics from exposure to radiation.

## Concerned Citizens for a Nuclear Breeder

P.O. Box 3
Ross, OH 45061
(513) 738-6750
*Harry Horner, Spokesperson*

Supports nuclear energy as a way to be independent of oil nations. Publishes *Newclips.*

## Congressional Alcohol Fuels Caucus

Office of Representative Richard Durbin
417 Cannon
Washington, DC 20515
(202) 225-5271
*Tom Faletti, Contact*

Members of Congress who support the development and use of alcohol fuels such as ethanol.

## Congressional Coal Group

343 Cannon House Office Building
Washington, DC 20515
(202) 226-7761
*Rep. Nick J. Rahall II, Chairman*

Bipartisan members of the House of Representatives interested in promoting coal as a fuel source.

## Conservation and Renewable Energy Inquiry and Referral Service

P.O. Box 8900
Silver Spring, MD 20907
(800) 523-2929
*Lawrence Hughes, Project Manager*

Acts as a clearinghouse for the U.S. Department of Energy to aid the transfer of technology through disseminating public information.

## Consumer Energy Council of America Research Foundation

2000 L Street, NW, Suite 802
Washington, DC 20036
(202) 659-0404
*Ellen Berman, Executive Director*

Advocates the consumers' interest in national energy policy before Congress and other public forums.

## Cosanti Foundation

6433 Doubletree Road
Scottsdale, AZ 85253
(602) 948-6145
*Paolo Soleri, President*

Main project is the building of the energy-efficient town of Arcosanti, an arcology (architecture and ecology) prototype.

## Council of Energy Resource Tribes

1580 Logan Street, Suite 400
Denver, CO 80203
(303) 832-6600
*David Lester, Executive Director*

American Indian tribes who own energy resources and promote conservation and control of their oil, coal, uranium, and natural gas.

## Council on Alternate Fuels

1225 I Street, NW, Suite 320
Washington, DC 20005
(202) 898-0711
*Michael S. Koleda, President*

Companies interested in the production of synthetic fuels.

## Critical Mass Energy Project
215 Pennsylvania Avenue, SE
Washington, DC
(202) 546-4996
*Ken Bossong, Director*

Created to stop the growth of nuclear power, make existing plants less dangerous, and promote alternatives.

## Energy Conservation Coalition
1525 New Hampshire Avenue, NW
Washington, DC 20036
(202) 745-4874
*Nick Fedoruk, Director*

Promotes energy efficiency as a means of meeting national energy needs.

## Energy Research Institute
6850 Rattlesnake Hammock Road
Naples, FL 33962
(813) 793-1922
*J. C. Caruthers, President and
Executive Director*

Individuals and companies interested in alternative energy sources.

## Environmental and Energy Study Institute
122 C Street, NW
Washington, DC 20001
(202) 628-1400
*Ken Murphy, Executive Director*

Educates national policy makers on energy vs. the environment.

## Environmental Coalition on Nuclear Power
433 Orlando Avenue
State College, PA 16803
(814) 237-3900
*Judith H. Johnsrud, Executive
Officer*

Assists in writing legislation regarding nuclear power plants and radioactive waste management.

## Exxon Corporation
1251 Avenue of the Americas
New York, NY 10020
(212) 333-6900
*Lawrence G. Rawl, CEO*

## Fusion Power Associates
Two Professional Drive, Suite 248
Gaithersburg, MD 20879
(301) 258-0545
*Stephen O. Dean, President*

Seeks to encourage and promote the development of fusion power as a viable energy option.

## GE Stockholders Alliance Against Nuclear Power & Weapons
P.O. Box 966
Columbia, MD 21044
(301) 381-2714
*Patricia T. Birnie, Chairman*

Objectives are to phase out GE's nuclear-related businesses and expand research of renewable energy resources.

## General Atomics
10955 John Jay Hopkins Drive
San Diego, CA 92121
(619) 455-3000
*Neil Blue, President*

Nuclear plant builder.

## General Electric (GE)
3135 Easton Turnpike
Fairfield, CT 06430
(203) 373-2211
*John F. Welch, Jr., Chairman*

## Get Oil Out
P.O. Box 1513
Santa Barbara, CA 93102
(805) 965-1519
*Robert Hopps, President*

Monitors and seeks to limit offshore gas and oil operations in the Santa Barbara Channel.

72 •   THE ENVIRONMENTAL ADDRESS BOOK

## Gulf Corporation
Post Office Box 1166
Pittsburgh, PA 15230
(412) 263-5000
*J. E. Lee, CEO*

## Health and Energy Institute
P.O. Box 5357
Takoma Park, MD 20912
(301) 585-5541
*Kathleen Tucker, President*

Research and education geared toward
a healthy environment and safe energy.

## Home Energy
2124 Kittredge Street, #95
Berkeley, CA 94704
(415) 524-5405

Magazine and consumer guide to energy-efficient living.

## Illinois Department of Energy and Natural Resources
325 W. Adams Street, Room 300
Springfield, IL 62704
(217) 785-2800
*Don Etchison, Director*

State agency that oversees energy and
natural resource issues.

## Institute for Resource and Security Studies
27 Ellsworth Avenue
Cambridge, MA 02139
(617) 491-5177
*Gordon Thompson, Executive Director*

Promotes nuclear and conventional
disarmament and environmental protection.

## Institute of Gas Technology
3424 South State Street
Chicago, IL 60616
(312) 567-3650
*Bernard S. Lee, President*

Promotes non-nuclear energy along
with the production and research of all
gas fuels.

## International Association for Hydrogen Energy
P.O. Box 24866
Coral Gables, FL 33124
(305) 284-4666
*T. Nejat Veziroglu, President*

Scientists and engineers involved in the
production and use of hydrogen.

## International Atomic Energy Agency
Vienna International Center
Wagramerstrasse 5, Postfach 100
A-1400 Vienna, Austria
222 2360
*H. Blix, Director General*

World government group that governs
and regulates the production and use of
nuclear energy.

## Interstate Oil Compact Commission
P.O. Box 53127
Oklahoma City, OK 73152
(405) 525-3556
*W. Timothy Dowd, Executive Director*

Organization of states that produce oil
and/or gas.

## Interstate Solar Coordination Council
900 American Center Building
St. Paul, MN 55101
(612) 296-4737
*John R. Dunlop, Chairman*

Managers of state government renewable energy programs.

## Jobs in Energy
1120 Riverside Avenue
Baltimore, MD 21230
(301) 659-0683
*Dennis Livingston, Executive Director*

Civil rights, religious, labor, environmental, and community organizations
who together create community-based
energy conservation enterprises providing jobs.

## Land Educational Associates Foundation

3368 Oak Avenue
Stevens Point, WI 54481
(715) 344-6158
*Gertrude Dixon, Director*

Researches nuclear power, waste, and weapons.

## League Against Nuclear Dangers

Rt. 1, 525 River Road
Rudolph, WI 54475
(715) 423-7996
*Naomi Jacobson, Co-Chairman and Director*

Educates the public on the dangers of nuclear power plant proliferation.

## Luz International

924 Westwood Boulevard,
Suite 1000
Los Angeles, CA 90024
(213) 208-7444
*Arnold Goldman, President*

One of the country's most successful solar power producers.

## Mobil Corporation

150 East 42nd Street
New York, NY 10017
(212) 883-4242
*Allen E. Murray, President*

## Mobilization for Survival

45 John Street, Suite 811
New York, NY 10038
(212) 385-2222
*Denise Wooden, Facilitator*

Advocates local action against nuclear power and military weapons facilities.

## National Association of Atomic Veterans

11004 East 40 Highway
Independence, MO 64055
(816) 737-9434
*Lucius Smith III, Adjutant General*

Veterans of U.S. nuclear weapons testing and Nagasaki/Hiroshima occupation forces who now have cancer believed to have been caused by exposure to radiation.

## National Association of Radiation Survivors

420 40th Street
Oakland, CA 94609
(415) 655-4886
*Fred Allingham, Administrative Director*

Uranium miners, test-site workers, civilian workers at national laboratories, atomic veterans, and down-wind residents working to provide health care and compensation.

## National Campaign for Radioactive Waste Safety

P.O. Box 4524, 105 Stanford, SE
Albuquerque, NM 87106
(505) 262-1862
*Peter Montague, Director*

Publishes *The Workbook*, which contains information on government handling of radioactive waste.

## National Coal Association

1130 17th Street, NW
Washington, DC 20036
(202) 463-2655
*Richard L. Lawson, President*

Producers who promote the use of coal as fuel.

## National Committee for Radiation Victims

6935 Laurel Avenue
Tacoma Park, MD 20912
(301) 891-3990
*E. Cooper Brown, Director*

Serves Americans affected by exposure to man-made radiation.

**National Consumer Research Institute**
2045 North 15th Street, Suite 309
Alexandria, VA 22201
(703) 528-5755
*Glenn H. Lovin, Treasurer*

Conducts energy research for federal, state, and local governments.

**National Energy Foundation**
5160 Wiley Post Way, Suite 200
Salt Lake City, UT 84116
(801) 539-1406
*Edward Dalton, President*

Works to stimulate interest and increase knowledge on the current energy situation through nationwide education programs for teachers and students.

**National Food and Energy Council**
409 Vandiver West, Suite 202
Columbia, MO 65202
(314) 875-7155
*Kenneth L. McFate, Executive Manager*

Electric industry group that works to inform and educate consumers about the critical interdependence of energy and food.

**National Hydropower Association**
1133 21st Street, NW, Suite 500
Washington, DC 20036
(202) 331-7551
*Elaine Evans, Executive Director*

Hydrodevelopers, dam site owners and manufacturers who promote hydroelectric energy.

**National Petroleum Council**
1625 K Street, NW, 6th Floor
Washington, DC 20006
(202) 393-6100
*Marshall W. Nichols, Executive Director*

Advisory council to the Secretary of Energy on matters relating to oil and gas.

**National Wood Energy Association**
P.O. Box 498
Pepperell, MA 01463
(617) 433-5674
*Scott Sklar, Executive Director*

Advocates wood as a renewable energy resource.

**Natural POWWER**
5420 Mayfield Road
Cleveland, OH 44124
(216) 442-5600
*Irwin Friedman, Chairman*

Advocates the replacement of nuclear power with alternative sources of energy.

**Nuclear Action Project**
2020 Pennsylvania Avenue, Suite 103
Washington, DC 20006
(202) 331-9831
*Rob Hager, Director*

Works to restore the rights of state and local governments to protect their citizens from health and safety hazards.

**Nuclear Free America**
325 East 25th Street
Baltimore, MD 21218
(301) 235-3575
*Albert Donnay, Executive Director*

Advises on how to create a Nuclear Free Zone in your community.

**Nuclear Information and Resource Service**
1424 16th Street, NW
Washington, DC 20036
(202) 328-0002
*Michael Mariotte, Executive Director*

Concerned with nuclear power issues; publishes *Groundswell*.

## Nuclear Recycling Consultants
P.O. Box 819
Provincetown, MA 02657
(508) 487-1930
*Jay M. Critchley, Director*

Promotes the conversion of nuclear facilities into shopping malls, museums, and condominiums, particularly the Seabrook plant into a nuclear energy monument.

## Nuclear Suppliers Association
c/o Baird Corporation
125 Middlesex Turnpike
Bedford, MA 01730
(617) 276-6204
*Stephen F. Grover, President*

Association composed of manufacturers and distributors of nuclear supplies.

## Occidental Petroleum
10889 Wilshire Boulevard
Los Angeles, CA 90024
(213) 879-1700
*Ray Irani, President*

## Organization of Petroleum Exporting Countries (OPEC)
Obere-Donau-Strasse 93
A-1020 Vienna, Austria
*Fadhil Al-Chalabi, Deputy Secretary General*

## Oxygenated Fuels Association
1330 Connecticut Avenue, NW, #300
Washington, DC 20036
(202) 822-6750
*George S. Dominguez, Executive Director*

Manufacturers and users of methanol and methyl-tertiary-butyl-ether for motor fuel.

## Passive Solar Industries Council
2836 Duke Street
Alexandria, VA 22314
(703) 823-3356
*Layne Ridley, Executive Director*

Seeks to provide an expanded commercial market for passive solar energy products and technology.

## Passive Solar Institute
P.O. Box 722
Bascom, OH 44809
(419) 937-2225
*Joseph Deahl, President*

Designers, engineers, and others promoting passive solar building designs.

## Pennsylvania Energy Office
P.O. Box 8010
Harrisburg, PA 17105
(717) 783-9981
*Mark S. Singel, Chairman*

The principal state agency for the development of energy policy.

## People for Energy Progress
P.O. Box 777
Los Gatos, CA 95031
(408) 925-2443
*K. T. Schaefer, Managing Director*

Promotes development and implementation of an energy policy which also improves the standard of living.

## People's Energy Resource Cooperative
354 Waverly Street
Framingham, MA 01701
(508) 879-8572

Sells caulks, glazing supplies, foams, tapes, weather strippings, etc., to make home heating more efficient.

## Phillips Petroleum Company
Phillips Building
Bartlesville, OK 74003
(918) 661-6000
*C. J. Silas, CEO*

## Polarized International
Box A
Tarzana, CA 91356
(818) 881-5525
*Myron Kahn, President*

Manufacturer and distributor of solar energy products.

## Radioactive Waste Campaign
625 Broadway, 2nd Floor
New York, NY 10012-2611
(212) 437-7390
*Minard Hamilton, Director*

Conducts research and public forums on the problems of radioactive waste disposal.

## Renew America
1001 Connecticut Avenue, NW, Suite 719
Washington, DC 20036
(202) 466-6880
*Tina Hobson, Executive Director*

Formerly known as Fund for Renewable Energy and the Environment. Individuals dedicated to the use of renewable energy and conservation.

## Renewable Energy Info Center
3201 Corte Maidas, Unit 4
Camarillo, CA 93010
(805) 388-3097
*David Foster, Executive Officer*

Studies all aspects of renewable energy. Formerly Geothermal World Info Center.

## Resources for the Future
1616 P Street, NW
Washington, DC 20036
(202) 328-5000
*Robert W. Fri, President*

Develops renewable resource projects and provides information to local groups.

## Rocky Mountain Institute
1739 Snowmass Creek Road
Snowmass, CO 81654
(303) 927-3851

An information source on super-insulated homes and other resource-efficient living.

## Safe Energy Communication Council
1717 Massachusetts Avenue, NW, Suite LL215
Washington, DC 20036
(202) 483-8491
*Scott Denman, Director*

Promotes the dissemination of fair and accurate energy information through broadcast and print media.

## Shell Oil Corporation
One Shell Plaza
Houston, TX 77001
(713) 241-6161
*L. C. van Wachem, Chairman*

## Siemans Solar
4530 Adohr Lane
Camarillo, CA 93010
(805) 482-6800
*Charles F. Gay, President*

World's largest maker of photovoltaic cells and solar modules.

## Solar Energy Industry Association
1730 North Lynn Street, Suite 610
Arlington, VA 22209
(703) 524-6100
*Scott Sklar, Executive Director*

Manufacturers, installers, distributors, contractors, and engineers of solar energy systems and components.

## Solartherm
1315 Apple Avenue
Silver Spring, MD 20910
(301) 587-8686
*Carl Schleicher, President*

Researches and develops low-cost solar energy systems.

## South Carolina Energy Office
P.O. Box 11405
Columbia, SC 29211
(803) 734-1740
*John F. Clark, Executive Director*

State government agency concerned with energy issues.

## Standard Oil Company
200 Public Square
Cleveland, OH 44114
(216) 586-4141
*R. B. Horton, CEO*

## Sunderland, J. Edward
University of Massachusetts
Amherst, MA 01003
(413) 545-0111

Solar energy scientist and pioneer.

## Task Force Against Nuclear Pollution
P.O. Box 1817
Washington, DC 20013
(301) 474-8311
*John W. Gofman, President*

Educates the public and influences the government on problems of nuclear pollution.

## Tennessee Valley Authority (TVA)
400 West Summit Hill Drive
Knoxville, TN 37902
(615) 632-2101
*C. H. Dean, Jr., Chairman of the Board*

Created in 1933 for the development of a region including Tennessee, Kentucky, Mississippi, Alabama, Virginia, Georgia, and North Carolina through the use of hydroelectric power.

## Texaco
2000 Westchester Avenue
White Plains, NY 10650
(914) 253-4000
*John K. McKinley, CEO*

## Texas Bureau of Economic Geology
University of Texas at Austin
University Station, Box X
Austin, TX 78713-7508
(512) 471-1534
*W. L. Fisher, Director*

State geological survey.

## Three Mile Island Alert
315 Peffer Street
Harrisburg, PA 17102
(717) 233-3072
*Kay Pickering, Coordinator*

Concerned about the potentially hazardous effects of nuclear reactors and seeks to close and clean up the reactor that leaked in 1979.

## United Nations Scientific Committee on the Effects of Atomic Radiation
Vienna International Centre, P.O. Box 500
A-1400 Vienna, Austria
26310

## U.S. Council for Energy Awareness
1776 I Street, NW, Suite 400
Washington, DC 20006
(202) 293-0770
*Harold B. Finger, CEO*

Electric utilities, manufacturers, industrial firms, research, and service organizations engaged in development of nuclear energy.

## U.S. Energy Association
1620 I Street, Suite 615
Washington, DC 20006
(202) 331-0415
*Barry K. Worthington, Executive Director*

Represents the energy interests of industry, government, professional and technical societies, educational institutions, legal organizations, etc.

## U.S. Nuclear Regulatory Commission (NRC)
1717 H Street, NW
Washington, DC 20555
(202) 492-7000
*Lando W. Zech, Jr., Chairman*

Regulates the civilian use of nuclear energy.

California Office
1450 Maria Lane
Walnut Creek, CA 94597
(415) 943-3809

## U.S. Windpower
6952 Preston Avenue
Livermore, CA 94550
(415) 455-6012

The largest American wind energy company.

## Unocal
Unocal Center
Los Angeles, CA 90017
(213) 977-7600
*Richard J. Stegemeier, CEO*

## Utah Department of Natural Resources
Energy Office
355 West North Temple, 3 Triad Center, Suite 450
Salt Lake City, UT 84180-1204
(801) 538-5428
*Richard M. Anderson, Director*

State government agency concerned with energy issues.

## Utility Nuclear Waste and Transportation Program
1111 19th Street, NW
Washington, DC 20036
(202) 778-6511
*Steven P. Kraft, Director*

Electric utilities united to promote solutions to the problem of radioactive waste management.

## Western Interstate Energy Board
6500 Stapleton Plaza, 3333 Quebec Street
Denver, CO 80207
(303) 377-9459
*Douglas C. Larson, Executive Director*

Purpose is to assist 16 western states in influencing the federal government on energy issues affecting the region.

## Westinghouse Electric
6 Gateway Center, Westinghouse Building
Pittsburgh, PA 15222
(412) 642-3800
*J. C. Marous, CEO*

## White Electric: The Lightbulb Place
1511 San Pablo Avenue
Berkeley, CA 94702
(800) 468-2852
(415) 548-2852

Mail-order firm specializing in energy-saving lightbulbs.

## Wood Heating Alliance
1101 Connecticut Avenue, NW, Suite 700
Washington, DC 20036
(202) 857-1181
*Carter Keithley, Executive Director*

Promotes research and education in the use of wood, coal, and other fuels as alternative sources of energy.

# GOVERNMENT

This chapter includes all of the major U.S. federal and state government agencies which have broad responsibilities regarding the environment. (For U.S. state agencies with a single concern, such as Departments of Agriculture, refer to the chapter covering that subject.) Also included here are all congressional committees dealing with any aspect of ecological issues, all national parks and some national seashores, private U.S. organizations whose major function is lobbying for legislation or whose interests are broad and statewide, and international groups with a wide range of concerns. Government and nongovernment agencies of countries other than the United States which have the greatest responsibilities for environmental protection are also found in this chapter. (Many nongovernment groups in Canada, however, are listed with their major subject of concern.)

## U.S. FEDERAL GOVERNMENT

**Bush, George Herbert Walker**
President of the United States of America
1600 Pennsylvania Avenue, NW
Washington, DC 20500

**U.S. Congress**
Office of Technology Assessment
600 Pennsylvania Avenue, SE
Washington, DC 20510
(202) 224-8713

Helps Congress plan for the environmental consequences of technology.

**U.S. Council on Environmental Quality**
722 Jackson Place, NW
Washington, DC 20503
(202) 395-5750
*Michael R. Deland, Chairman*
*David B. Struhs, Staff Director*

Assists and advises the President on conditions, trends, and the quality of the environment.

**U.S. Department of Agriculture**
14th Street and Independence Avenue, SW
Washington, DC 20250
(202) 447-2791
*Edward R. Madigan, Secretary*

Federal government agency responsible for regulating both the quality and the quantity of food produced in the United States.

Agricultural Stabilization and Conservation Service
P.O. Box 2415
Washington, DC 20013
(202) 447-5237
*Keith D. Bjerke, Administrator*
*James R. McMullen, Director of Conservation and Environmental Protection*

Operates the Conservation
Reserve, Agricultural
Conservation, and Clean
Water programs.

Animal and Plant Health
Inspection Service
14th Street and Independence
Avenue, SW
Washington, DC 20250
(301) 436-7779
*James W. Glosser, Administrator*
*Larry B. Slagle, Associate*
*Administrator*

Economic Research Service
1301 New York Avenue, NW,
Room 1226
Washington, DC 20005-4788
(202) 786-1504
*John E. Lee, Jr., Administrator*
Studies the economic effect of
environmental quality
improvement.

Forest Service
P.O. Box 2417
Washington, DC 20013
(202) 447-3760
*F. Dale Robertson, Chief*
*George M. Leonard, Associate*
*Chief*

• Forest Service—Region 1,
Northern
Federal Building, P.O. Box
7669
Missoula, MT 59807
(406) 329-3326
*Kirk M. Horn, Director of*
*Wildlife and Fisheries*

• Forest Service—Region 2,
Rocky Mountain
11177 West 8th Avenue, Box
25127
Lakewood, CO 80255
(303) 236-9427
*Gary E. Cargill, Regional*
*Forester*

• Forest Service—Region 3,
Southwestern
Federal Building, 517 Gold
Avenue, SW
Albuquerque, NM 87102
(505) 476-3300
*David F. Jolly, Regional*
*Forester*

• Forest Service—Region 4,
Intermountain
Federal Office Building, 324
25th Street
Ogden, UT 84401
(801) 625-5605
*Stan Tixier, Regional Forester*

• Forest Service—Region 5,
California
630 Sansome Street
San Francisco, CA 94111
(415) 556-4310
*Paul F. Barker, Regional*
*Forester*

• Forest Service—Region 6,
Pacific Northwest
319 SW Pine Street, Box 3623
Portland, OR 97208
(503) 326-3625
*James F. Torrence, Regional*
*Forester*

• Forest Service—Region 8,
Southern
1720 Peachtree Road, NW,
Suite 800
Atlanta, GA 30367
(404) 347-4177
*John E. Alcock, Regional*
*Forester*

• Forest Service—Region 9,
Eastern
310 Wisconsin Avenue
Milwaukee, WI 53203
(414) 291-3693
*Floyd J. Marita, Regional*
*Forester*

- Forest Service—Region 10, Alaska
Federal Office Building, Box 21628
Juneau, AK 99802-1628
(907) 586-8863
*Michael A. Barton, Regional Forester*

Natural Resources and Environment
14th Street and Independence Avenue, SW
Washington, DC 20250
(202) 447-4581
*Roland R. Vatour, Assistant Secretary*

Plant Protection and Quarantine
Federal Center Building
Hyattsville, MD 20782
(202) 447-5601
*William F. Helms, Deputy Administrator*
*Richard R. Backus, Assistant Deputy Administrator*

- Latin American Region
American Embassy, Reforma 395
Col. Cuahtemoc, 06506, Mexico City, D.F., Mexico
211-0042 3480
*Ed L. Ayers, Jr., Regional Director*

Science and Education Administration
14th Street and Independence Avenue, SW
Washington, DC 20250
(202) 447-8732
*Charles Hess, Assistant Secretary*

Soil Conservation
P.O. Box 2890
Washington, DC 20013
(202) 447-4543
*R. Mack Gray, Chief*

Veterinary Services
Federal Center Building
Hyattsville, MD 20782
(202) 447-5193
*Billy G. Johnson, Deputy Administrator*
Animal pests and diseases.

- Southeastern Region
700 Twiggs Street
Tampa, FL 33601
(813) 228-2952
*L. D. Konya, Regional Director*

- Central Region
Fort Worth, TX 76102
(817) 885-5566
*A. L. Strating, Regional Director*

- Northern Region
Building 2, GSA Depot
Scotia, NY 12302
(518) 370-5026
*George P. Pierson, Regional Director*

- Western Region
P.O. Box 3857, 317 Inverness Way South
Englewood, CO 80112
(303) 796-6851
*R. Harrington, Jr., Regional Director*

## U.S. Department of Commerce

National Atmospheric and Oceanic Administration
14th and Constitution Avenues, Room 5128
Washington, DC 20230
(202) 377-2985
*Undersecretary (Vacant)*
*B. Kent Burton, Assistant Secretary and Deputy Administrator*

Responsible for all commercial and marine fisheries within the 200-nautical-mile limit, environmental satellite research, etc.

## U.S. Department of Defense
The Pentagon
Washington, DC 20301
(703) 545-6700
*Richard Cheney, Secretary of Defense*

Responsible for nuclear weapon production.

Army Corps of Engineers
Office of the Chief of Engineers, Pulaski Building
200 Massachusetts Avenue, NW
Washington, DC 20314
(202) 272-0001
*Henry J. Hatch, Chief of Engineers*
Conducts civil works missions to maintain and create conditions in which human and natural environments can exist in productive harmony.

## U.S. Department of Education
Office of Environmental Education
400 Maryland Avenue, SW
Washington, DC 20202
(202) 708-5366

## U.S. Department of Energy
1000 Independence Avenue, SW
Washington, DC 20585
(202) 252-5000
*James D. Watkins, Secretary of Energy*
*Michael R. McElwrath, Assistant Secretary, Fossil Energy; John R. Berg, Assistant Secretary, Conservation and Renewable Energy; Peter N. Brush, Assistant Secretary, Environment, Safety and Health; Jerry O. Griffith, Assistant Secretary, Nuclear Energy.*

Energy Information Administration
Forrestal Building, 1000 Independence Avenue, SW
Washington, DC 20585
(202) 586-6210
*Helmut A. Merklein, Administrator*
Central energy data collection and analysis arm of the federal government.

## U.S. Department of Health and Human Services
Centers for Disease Control (CDC)
1600 Cliffton Road, NE
Atlanta, Georgia 30333
(404) 639-3311
*William Roper, Director*

Studies disease in all its facets, including possible links between the environment and cancer.

National Institute of Environmental Health Sciences
P.O. Box 12233
Research Triangle Park, NC 27709
(919) 541-2111
*James Mason, Assistant Secretary*

## U.S. Department of the Interior
1849 C Street, NW
Washington, DC 20240
(202) 208-5048
*Manuel Lujan, Jr., Secretary*

The major U.S. government agency responsible for fish and wildlife, land management, mining, national parks and forests, and Indian lands.

Bureau of Land Management
C Street between 18th and 19th Streets, NW
Washington, DC 20240
(202) 343-9435
*Cy Jamison, Director*
Manages federal public lands.

Fish and Wildlife Service
C Street between 18th and 19th
Streets, NW
Washington, DC 20240
(202) 343-5634
*John F. Turner, Director*

• Division of Endangered Species
and Habitat Conservation
C Street between 18th and
19th Streets, NW
Washington, DC 20240

• Office of Environmental
Service
300 Ala Moana Boulevard,
P.O. Box 50167
Honolulu, HI 96850

National Park Service
C Street, between 18th and 19th
Streets, NW
Washington, DC 20240
(202) 208-6843
*James Ridenour, Director*

• Acadia National Park
P.O. Box 177
Bar Harbor, ME 04609
(207) 288-3338
*John A. Hauptman,*
*Superintendant*

• American Samoa National
Park
c/o Pacific Area Director
P.O. Box 50165
Honolulu, HI 96850

• Arches National Park
446 South Main Street
Moab, UT 84532
(801) 259-8161
*Paul D. Guraedy,*
*Superintendant*

• Assateague Island National
Seashore
Rt. 2, Box 294
Berlin, MD 21811
(301) 641-1441
*Roger K. Rector,*
*Superintendant*

• Badlands National Park
P.O. Box 6
Interior, SD 57750
(605) 433-5361
*Irvin L. Mortenson,*
*Superintendant*

• Big Bend National Park
Big Bend National Park, TX
79834
(915) 477-2251
*James W. Carrico,*
*Superintendant*

• Biscayne National Park
P.O. Box 1369
Homestead, FL 33030
(305) 247-2044
*James A. Sanders,*
*Superintendant*

• Bryce Canyon National Park
Bryce Canyon, UT 84717
(801) 834-5322
*Robert W. Reynolds,*
*Superintendant*

• Canaveral National Seashore
P.O. Box 6447
Titusville, FL 32782-8447
(305) 867-0634

• Canyonlands National Park
446 South Main Street
Moab, UT 84532
(801) 259-7165
*Harvey D. Wickware,*
*Superintendant*

• Cape Cod National Seashore
South Wellfleet, MA 02663
(617) 349-3785
*Herbert Olsen, Superintendant*

• Cape Hatteras National
Seashore
Rt. 1, Box 695
Manteo, NC 27964
(919) 473-2111

• Cape Lookout National
Seashore
P.O. Box 690
Beaufort, NC 28516
(919) 728-2121
*William A. Harris,*
*Superintendant*

• Capitol Reef National Park
Torrey, UT 84775
(801) 425-3871
*Martin C. Ott, Superintendant*

• Carlsbad Caverns National
Park
3225 National Parks Highway
Carlsbad, NM 88220
(505) 885-8884
*Wallace B. Elms,
Superintendant*

• Channel Islands National Park
1801 Spinnaker Drive
Ventura, CA 93001
(805) 644-8157

• Crater Lake National Park
P.O. Box 7
Crater Lake, OR 97604
(503) 594-2211
*Robert E. Benton,
Superintendant*

• Cumberland Island National
Seashore
P.O. Box 806
Saint Mary's, GA 31558
(912) 882-4336

• Denali National Park
P.O. Box 9
McKinley Park, AK 99755
(907) 883-2284
*Russell W. Berry, Jr.,
Superintendant*

• Everglades National Park
P.O. Box 279
Homestead, FL 33030
(305) 247-6211
*Michael V. Finley,
Superintendant*

• Fire Island National Seashore
120 Lalure Street
Patchogue, NY 11772
(516) 289-4810
*Noel J. Pachta, Superintendant*

• Gates of the Arctic National
Park
P.O. Box 74680
Fairbanks, AK 99707
(907) 452-5363
*Roger J. Siglin, Superintendant*

• Glacier Bay National Park
Gustavus, AK 99826
(907) 697-2232
*Marvin Jensen, Superintendant*

• Glacier National Park
West Glacier, MT 59936
(406) 888-5441
*H. Gilbert Lusk,
Superintendant*

• Grand Canyon National Park
P.O. Box 129
Grand Canyon, AZ 88023
(802) 638-7888
*John H. David, Superintendant*

• Grand Teton National Park
P.O. Drawer 170
Moose, WY 83012
(307) 733-2880
*Jack E. Stark, Superintendant*

• Great Basin National Park
Baker, NV 89311
(702) 234-7331
*Albert J. Hendricks,
Superintendant*

• Great Smoky Mountains
National Park
Gatlinburg, TN 37738
(615) 436-5615
*Randall R. Pope,
Superintendant*

• Guadalupe Mountains
National Park
HC 60, Box 400
Salt Flat, TX 79847-9400
(915) 828-3351
*Karen Wade, Superintendant*

• Gulf Islands National Seashore
P.O. Box 100
Gulf Breeze, FL 32561
(904) 932-5302
*Jerry A. Eubanks,
Superintendant*

- Haleakala National Park
P.O. Box 369
Makawao, Maui, HI 96768
(808) 572-9177
*Donald W. Reeser,*
*Superintendant*

- Hawaii Volcanoes National
Park
Hawaii National Park, HI
96718
(808) 967-7311
*Hugo H. Huntzinger,*
*Superintendant*

- Hot Springs National Park
P.O. Box 1880
Hot Springs, AR 71801
(501) 624-3383
*Roger Giddings,*
*Superintendant*

- Isle Royale National Park
87 North Ripley Street
Houghton, MI 49931
(906) 482-3310
*Thomas C. Hobbs,*
*Superintendant*

- Katmai National Park
P.O. Box 7
King Salmon, AK 99613
(907) 246-3305
*Ray Bane, Superintendant*

- Kenai Fjords National Park
P.O. Box 1727
Seward, AK 99664
(907) 224-3874
*Anne Castellina,*
*Superintendant*

- Kobuk Valley National Park
P.O. Box 287
Kotzebue, AK 99752
(907) 442-3890
*Alan D. Eliason,*
*Superintendant*

- Lake Clark National Park
P.O. Box 61
Anchorage, AK 99513
(907) 781-2205
*Andrew Hutchison,*
*Superintendant*

- Lassen Volcanic National Park
Mineral, CA 96063
(916) 595-4444
*Gilbert E. Blinn,*
*Superintendant*

- Mammoth Cave National Park
Mammoth Cave, KY 42259
(502) 758-2251

- Mesa Verde National Park
Mesa Verde National Park,
CO 81330
(303) 529-4465
*Robert C. Heyder,*
*Superintendant*

- Mount Rainier National Park
Tacoma Woods, Star Route
Ashford, WA 98304
(206) 569-2211
*Neal G. Guse, Superintendant*

- North Cascades National Park
800 State Street
Sedro Woolley, WA 98284
(206) 855-1331
*John J. Earst, Superintendant*

- Olympic National Park
600 East Park Avenue
Port Angeles, WA 98362
(206) 452-4501
*Robert S. Chandler,*
*Superintendant*

- Padre Island National Seashore
9405 South Padre Island Drive
Corpus Christi, TX 78418
(512) 937-2621
*John D. Hunter, Jr.,*
*Superintendant*

- Petrified Forest National Park
Petrified Forest National Park,
AZ 86028
(602) 524-6228

- Point Reyes National Seashore
Point Reyes, CA 94956
(415) 663-8522
*John L. Sansing,*
*Superintendant*

• Redwood National Park
1111 Second Street
Crescent City, CA 95531
(707) 464-6101
*William H. Ehorn,
Superintendant*

• Rocky Mountain National
Park
Estes Park, CO 80517
(303) 586-2371
*James B. Thompson,
Superintendant*

• Sequoia and Kings Canyon
National Parks
Three Rivers, CA 93271
(209) 565-3341
*J. Thomas Ritter,
Superintendant*

• Shenandoah National Park
Luray, VA 22835
(703) 999-2243
*William Wade, Superintendant*

• Theodore Roosevelt National
Park
P.O. Box 7
Medora, ND 58645
(701) 623-4466
*Charles M. Shaver,
Superintendant*

• Virgin Islands National Park
P.O. Box 7789
Charlotte Amalie, St. Thomas,
Virgin Islands 00801
(809) 775-2050
*Marc A. Koenings,
Superintendant*

• Voyageurs National Park
P.O. Box 50
International Falls, MN 58649
(216) 283-9821
*Benton K. Clary,
Superintendant*

• Wind Cave National Park
Hot Springs, SD 57747
(605) 745-4600
*Ernest W. Ortega,
Superintendant*

• Wrangell-St. Elias National
Park
P.O. Box 29
Glenn Allen, AK 99588
(907) 822-5235
*Richard H. Martin,
Superintendant*

• Yellowstone National Park
P.O. Box 168
Yellowstone National Park,
WY 82190
(307) 344-7381
*Robert D. Barbee,
Superintendant*

• Yosemite National Park
P.O. Box 577
Yosemite National Park, CA
95389
(208) 372-4461
*B. J. Griffin, Superintendant*
(Acting)

• Zion National Park
Springdale, UT 84767
(801) 772-3256
*Harold L. Grafe,
Superintendant*

## U.S. Department of Justice

Environment and Natural
Resources Division
10th Street and Constitution
Avenue, NW
Washington, DC 20530
(202) 514-2701
*Richard B. Stewart, Assistant
Attorney General*

Prosecuting agency for offenders of federal environmental protection statutes.

## U.S. Department of Labor

Occupational Safety and Health
Administration
200 Constitution Avenue, NW
Washington, DC 20210
(202) 523-8017
*Gerard F. Scannell, Assistant
Secretary*

Government division concerned with industrial accidents, including toxic and nuclear.

## U.S. Department of Science and Education

Agricultural Research
S&E, Room 217-W
Washington, DC 20250
(202) 447-3656
*R. D. Plowman, Administrator*

Conducts research in watershed engineering, soil management, water management, and air pollution control.

## U.S. Department of Transportation

400 7th Street, SW
Washington, DC 20590
(202) 426-4000
*Samuel K. Skinner, Secretary of Transportation*

Federal agency responsible for highway construction, the safety of vehicles, and other issues that impact on our air, land, and water quality.

## U.S. Department of Veteran's Affairs

810 Vermont Avenue, NW
Washington, DC 20420
(202) 233-4000
*Edward J. Derwinski, Secretary*

Responsible for the medical care of veterans affected with chemical and biological warfare agents.

## U.S. Environmental Protection Agency (EPA)

401 M Street, SW
Washington, DC 20460
(202) 797-6829
(800) 535-0202 Hazardous chemicals hotline
(800) 426-4791 Safe drinking water hotline
*William K. Reilly, Administrator*

U.S. government agency responsible for the environment.

Air Division, Office of Air and Radiation
401 M Street, SW
Washington, DC 20460
(202) 475-8470
Publishes *The Inside Story: A Guide to Indoor Air Quality,* focusing on pollution in the home, including radon.

Region 1
JFK Federal Building
Boston, MA 02203
(617) 565-3234
*Julie D. DeLaga, Administrator*

Region 2
26 Federal Plaza
New York, NY 10278
(212) 264-4418
*Constantine Sidamon-Eristoff, Administrator*

Region 3
841 Chestnut Street
Philadelphia, PA 19107
(215) 597-4084
*Edwin B. Erickson, Administrator*

Region 4
345 Courtland Street, NE
Atlanta, GA 30365
(404) 347-2904
*Greer C. Tidwell, Administrator*

Region 5
230 South Dearborn Street
Chicago, IL 60604
(312) 886-6165
*Valdas V. Adamkus, Administrator*

Region 6
1445 Ross Avenue
Dallas, TX 75202
(214) 655-7208
*Robert E. Layton, Administrator*

Region 7
726 Minnesota Avenue
Kansas City, KS 66101
(913) 236-2893
*Morris Kay, Administrator*

Region 8
999 18th Street, One Denver
Place, Suite 1300
Denver, CO 80202
(303) 293-1648
*James Scherer, Administrator*

Region 9
215 Fremont Street
San Francisco, CA 94105
(415) 974-8378
*Daniel McGovern, Administrator*

Region 10
1200 Sixth Avenue
Seattle, WA 98101
(206) 442-7660
*Robie G. Russell, Administrator*

University of Tennessee
328 South Stadium Hall, Room
429
Knoxville, TN 37996
(615) 974-1878
*John R. Paulk, Project Director*
Project headquarters for the
EPA's national network for
environmental education.

## U.S. Federal Trade Commission (FTC)

Bureau of Consumer Protection
Pennsylvania at Sixth Street, NW
Washington, DC 20580
(202) 326-2222
*Barry J. Cutler, Director*

Government agency responsible for
fair trade practices, including proper
labeling of food.

## U.S. Food and Drug Administration (FDA)

5600 Fishers Lane
Rockville, MD 20857
(301) 443-1544
*David A. Kessler, Commissioner*

U.S. agency responsible for the health
and safety of food products, including
chemical additives.

## U.S. General Services Administration

National Audiovisual Center
National Archives and Records
Service
Washington, DC 20409
(301) 763-1896

Handles audiovisual materials produced by or for the U.S. government.

## U.S. Geological Survey

Reston, VA 22092
(703) 648-4460
*Dallas L. Peck, Director*

Maps the geological features and resources of the United States.

## U.S. Government Printing Office

732 North Capitol Street, NW
Washington, DC 20402
(202) 275-3648

Prints innumerable pamphlets and reports on environmental topics.

# CONGRESSIONAL COMMITTEES

## U.S. Joint Committee

### Environmental and Energy Study Conference

House Annex II, Room 515
Washington, DC 20515
(202) 226-3300
*Rep. Bob Wise (WV), Co-Chairman*
*Sen. Albert Gore, Jr. (TN),*
*Co-Chairman*

Provides analysis of environmental, energy, and natural resources issues
before Congress.

## U.S. House of Representatives

### Committee on Agriculture
Room 1301, Longworth House Office Building
Washington, DC 20515
(202) 225-2171
*E. (Kika) de la Garza (TX), Chairman*

Responsible for protection of birds and animals in forest reserves, agricultural and industrial chemistry, soil conservation, and forestry in general.
*Majority* (Democrats): Walter B. Jones (NC), George E. Brown, Jr. (CA), Charles Rose (NC), Glenn English (OK), Leon E. Panetta (CA), Jerry Huckaby (LA), Dan Glickman (KS), Charles W. Stenholm (TX), Harold L. Volkmer (MO), Charles Hatcher (GA), Robin Tallon (SC), Halrey O. Staggers, Jr. (WV), David R. Nagle (IA), Jim Jontz (IN), Tim Johnson (SD), Claude Harris (AL), Ben Nighthorse Campbell (CO), Mike Espy (MS), Bill Sarpalius (TX), Jill Long (IN), Gary Condir (CA), Roy Dyson (MD), Martin Lancaster (NC).
*Minority* (Republicans): E. Thomas Coleman (MO), Ron Marlenee (MT), Larry J. Hopkins (KY), Arlen Stangeland (MN), Pat Roberts (KS), Bill Emerson (MO), Sid Morrison (WA), Steve Gunderson (WI), Tom Lewis (FL), Robert F. Smith (OR), Larry Combest (TX), Bill Schuette (MI), Fred Grandy (IA), Wally Herger (CA), Clyde C. Holloway (LA), James T. Walsh (NY), Bill Grant (FL).

Subcommittee on Conservation, Credit, and Rural Development
*Glenn English (OK), Chairman*

Subcommittee on Domestic Marketing, Consumer Relations, and Nutrition
*Charles Hatcher (GA), Chairman*

Subcommittee on Forests, Family Farms, and Energy
*Harold L. Volkmer (MO), Chairman*

### Committee on Energy and Commerce
Room 2125, Rayburn House Office Building
Washington, DC 20515
(202) 225-2927
*John D. Dingell (MI), Chairman*
*John Orlando, Chief of Staff*

Responsible for national energy policy, fossil fuels, nuclear and solar energy.
*Majority* (Democrats): James H. Scheuer (NY), Henry A. Waxman (CA), Philip R. Sharp (IN), Edward J. Markey (MA), Thomas A. Luken (OH), Doug Walgren (PA), Al Swift (WA), Cardiss Collins (IL), Mike Synar (OK), W.J. Tauzin (LA), Ron Wayden (OR), Ralph M. Hall (TX), Dennis E. Eckart (OH), Bill Richardson (NM), Jim Slattery (KS), Gerry Sikorski (FL), John Bryant (TX), Jim Bates (CA), Rick Boucher (VA), Jim Cooper (TN), Terry L. Bruce (IL), J. Roy Rowland (GA), Thomas J. Manton (NY), Edolphus Towns (NY), Tom McMillen (MD).
*Minority* (Republicans): Norman F. Lent (NY), Edward R. Madigan (IL), Carlos J. Moorhead (CA), Matthew J. Rinaldo (NJ), William E. Dannemeyer (CA), Bob Wittaker (KS), Thomas J. Tauke (IA), Don Ritter (PA), Thomas J. Billey, Jr. (VA), Jack Fields (TX), Michael G. Oxley (OH), Howard C. Nielson (UT), Michael Bilirakis (FL), Dan Schaefer (CO), Joe Barton (TX), Sonny Callahan (AL), Alex McMillan (NC).

Subcommittee on Energy and Power
*Philip R. Sharp (IN), Chairman*

Subcommittee on Health and the Environment
*Henry A. Waxman (CA), Chairman*

Subcommittee on Transportation and Hazardous Materials
*Thomas A. Luken (OH), Chairman*

## Committee on Interior and Insular Affairs
Room 1324, Longworth House Office Building
Washington, DC 20515
(202) 225-2761
*Morris K. Udall (AZ), Chairman*

Responsible for forest reserves, national parks, irrigation projects and Indian lands.
*Majority* (Democrats): George Miller (CA), Philip R. Sharp (IN), Edward J. Markey (MA), Austin J. Murphy (PA), Nick J. Rahall II (WV), Bruce F. Vento (MN), Pat Williams (MT), Beverly B. Byron (MD), Ron de Lugo (VI), Sam Gejdenson (CT), Peter H. Kostmayer (PA), Richard Lehman (CA), Bill Richardson (NM), George W. Darden (GA), Peter J. Visclosky (IN), Jaime B. Fuster (PR), Mel Levine (CA), James McClure Clarke (NC), Wayne Owens (UT), John Lewis (GA), Ben Nighthorse Campbell (CO), Peter A. DeFazio (OR), Eni F. H. Faleomavaega (AS), Jim McDermott (WA), Tim Johnson (SD).
*Minority* (Republicans): Don Young (AK), Robert J. Lagomarsino (CA), Ron Marlenee (MT), Larry E. Craig (ID), Denny Smith (OR), James V. Hansen (UT), Barbara F. Vucanovich (NV), Ben Garrido Blaz (Guam), John J. Rhodes III (AZ), Elton Gallegly (CA), Stan Parris (VA), Robert F. Smith (OR), Jim Ross Lightfoot (IA), Craig Thomas (WY), John F. Duncan, Jr. (TN).

Subcommittee on Energy and the Environment
*Morris K. Udall (AZ), Chairman*

Subcommittee on Mining and Natural Resources
*Nick J. Rahall II (WV), Chairman*

Subcommittee on National Parks and Public Lands
*Bruce F. Vento (MN), Chairman*

Subcommittee on Water, Power, and Offshore Energy Resources
*George Miller (CA), Chairman*

## Committee on Merchant Marine and Fisheries
Room 1334, Longworth House Office Building
Washington, DC 20515
(202) 225-4047
*Walter B. Jones (NC), Chairman*

Responsible for coastal zone management, international fishery agreements, etc.
*Majority* (Democrats): Walter B. Jones (NC), Gerry E. Studds (MA), Carroll Hubbard, Jr. (KY), William J. Hughes (NJ), Earl J. Hutto (FL), W.J. Tauzin (LA), Thomas M. Foglietta (PA), Dennis M. Hertel (MI), Roy Dyson (MD), William O. Lipinski (IL), Robert A. Borski (PA), Thomas R. Carper (DE), Douglas H. Bosco (CA), Robin Tallon (SC), Solomon P. Ortiz (TX), Charles E. Bennett (FL), Thomas J. Manton (NY), Owen B. Pickett (VA), Joseph E. Brennan (ME), George J. Hochbrueckner (NY), Bob Clement (TN), Stephen J. Solarz (NY), Frank Pallone, Jr. (NJ), Greg Laughlin (TX), Nita M. Lowey (NY), Jolene Unsoeld (WA), Gene Taylor (MS).
*Minority* (Republicans): Robert W. Davis (MI), Don Young (AK), Norman F. Lent (NY), Norman D. Shumway (CA), Jack Fields (TX), Claudine Schneider (RI), Herbert H. Bateman (VA), Jim Saxton (NJ), John Miller (WA), Helen Delich Bentley (MD), Howard Coble (NC), Curt Weldon (PA), Patricia Saiki (HI), Wally Herger (CA), Jim Bunning (KY), James M. Inhofe (OK), Porter J. Goss (FL), Arthur Ravenel, Jr. (SC).

Subcommittee on Fisheries and
Wildlife Conservation and the
Environment
*Gerry E. Studds (MA),
Chairman*

Subcommittee on Oceanography
and Great Lakes
*Dennis M. Hertel (MI),
Chairman*

Subcommittee on Surface
Transportation
*Norman Y. Mineta (CA),
Chairman*

Subcommittee on Water
Resources
*Henry J. Nowak (NY),
Chairman*

## Committee on Public Works and Transportation

Room 2165, Rayburn House Office
Building
Washington, DC 20512
(202) 225-4472
*Robert A. Roe (NJ), Chairman*

Responsible for flood control, improve-
ment of rivers and harbors, oil spills
and other pollution of navigable wa-
ters.
*Majority* (Democrats): Glenn Ander-
son (CA), Norman Y. Mineta (CA),
James L. Oberstar (MN), Henry J.
Nowak (NY), Nick J. Rahall II (WV),
Douglas Applegate (OH), Ron de Lugo
(VI), Gus Savage (IL), Douglas H.
Bosco (CA), Robert A. Borski (PA),
Joe Kolter (PA), Tim Valentine (NC),
William O. Lipinski (IL), Peter J. Vis-
closky (IN), James A. Traficant, Jr.
(OH), John Lewis (GA), Peter A.
DeFazio (OR), David E. Skaggs (CO),
James A. Hayes (LA), Bob Clement
(TN), Lewis F. Payne, Jr. (VA), Jerry
F. Costello (IL), Frank Pallone, Jr.
(NJ), Ben Jones (GA), Mike Parker
(MS), Greg Laughlin (TX), Pete Geren
(TX), George E. Sangmeister (IL),
Gary L. Ackerman (NY), Carl C. Per-
kins (KY).
*Minority* (Republicans): John P. Ham-
merschmidt (AR), Bud Shuster (PA),
Arlan Stangeland (MN), William F.
Clinger, Jr. (PA), Bob McEwen (OH),
Thomas E. Petri (WI), Ron Packard
(CA), Sherwood L. Boehlert (NY), Jim
Ross Lightfoot (IA), J. Dennis Hastert
(IL), James M. Inhofe (OK), Cass Bal-
lenger (NC), Fred Upton (MI), Bill
Emerson (MO), Bill Grant (FL), Larry
Craig (ID), John J. Duncan, Jr. (TN),
Melton D. Hancock (MO), Christo-
pher Cox (CA).

## Committee on Science, Space and Technology

Room 2181, Rayburn House Office
Building
Washington, DC 20515
(202) 225-2321
*Robert A. Roe (NJ), Chairman*

Responsible for NASA, National Sci-
ence Foundation.
*Majority* (Democrats): Robert A. Roe
(NJ), George E. Brown, Jr. (CA),
James H. Scheuer (NY), Marilyn
Lloyd (TN), Doug Walgren (PA), Dan
Glickman (KS), Harold L. Volkmer
(MO), Howard Wolpe (MI), Bill Nel-
son (FL), Ralph M. Hall (TX), Dave
McCurdy (OK), Norman Mineta
(CA), Tim Valentine (NC), Robert G.
Torricelli (NJ), Rick Boucher (VA),
Terry L. Bruce (IL), Richard Stallings
(ID), James A. Traficant, Jr. (OH), Lee
H. Hamilton (IN), Henry J. Nowak
(NY), Carl C. Perkins (KY), Tom
McMillen (MD), David E. Price (NC),
David R. Nagle (IA), James A. Hayes
(LA), David E. Skaggs (CO), Jerry F.
Costello (IL), Harry Johnston (FL),
John S. Tanner (TN), Glen Browder
(AL).
*Minority* (Republicans): Robert S.
Walker (PA), F. James Sensenbrenner
(WI), Claudine Schneider (RI), Sher-
wood L. Boehlert (NY), Tom Lewis
(FL) Don Ritter (PA), Sid Morrison
(WA), Ron Packard (CA), Paul B.
Henry (MI), Harris W. Fawell (IL), D.
French Slaughter, Jr. (VA), Lamar
Smith (TX), Jack Buechner (MO),
Constance A. Morella (MD), Christo-
pher Shays (CT), Dana Rohrabacher
(CA), Steven H. Schiff (NM), Tom J.
Campbell (CA), John J. Rhodes III
(AZ).

Subcommittee on Energy
Research and Development
*Marilyn Lloyd (TN), Chairman*

Subcommittee on Natural
Resources, Agriculture
Research, and Environment
*James H. Scheuer (NY),
Chairman*

Subcommittee on Transportation,
Aviation, and Materials
*Robert G. Torricelli (NJ),
Chairman*

## U.S. Senate

### Committee on Agriculture, Nutrition, and Forestry

Room 328-A, Russell Building
Washington, DC 20510
(202) 224-2035
*Patrick Leahy (VT), Chairman*
*Richard Lugar (IN), Ranking
Minority Member*

Committee on forestry, agriculture, agricultural pests, pesticides.
*Majority* (Democrats): David H. Pryor (AR), David L. Boren (OK), Howell Heflin (AL), Tom Harkin (IA), Kent Conrad (ND), Wyche Fowler, Jr. (GA), Thomas A. Daschle (SD), Max S. Baucus (MT), Bob Kerrey (NE).
*Minority* (Republicans): Jesse A. Helms (NC), Robert J. Dole (KS), Thad Cochran (MS), Rudy Boschwitz (MN), Mitch McConnell (KY), Christopher S. Bond (MO), Slade Gorton (WA).

Subcommittee on Conservation
and Forestry
*Wyche Fowler, Jr. (GA),
Chairman*

Subcommittee on Nutrition and
Investigations
*Tom Harkin (IA), Chairman*

### Committee on Commerce, Science, and Transportation

U.S. Senate—SD-508
Washington, DC 20510
(202) 224-5115
*Ernest F. Hollings (SC), Chairman*

Responsible for waterways, coastal management, oceans, and marine fisheries.
*Majority* (Democrats): Daniel K. Inouye (HI), Wendell H. Ford (KY), J. James Exon (NE), Albert Gore, Jr. (TN), John D. Rockefeller IV (WV), Lloyd Bentsen (TX), John F. Kerry (MA), John B. Breaux (LA), Richard H. Bryan (NV), Charles S. Robb (VA).
*Minority* (Republicans): John C. Danforth (MO), Bob Packwood (OR), Larry Pressler (SD), Ted Stevens (AK), Robert W. Kasten, Jr. (WI), John McCain (AZ), Conrad Burns (MT), Slade Gorton (WA), Trent Lott (MS).

### Committee on Energy and Natural Resources

Room SD-364, Dirksen Building
Washington, DC 20510
(202) 224-4971, 224-6163 FAX
*J. Bennett Johnston (LA),
Chairman*
*James A. McClure (ID), Ranking
Minority Member*

Responsible for regulation, conservation, and extraction of minerals, petroleum, and nuclear energy; also solar energy.
*Majority* (Democrats): J. Bennett Johnston (LA), Dale L. Bumpers (AR), Wendell H. Ford (KY), Howard M. Metzenbaum (OH), Bill Bradley (NJ), Jeff Bingaman (NM), Timothy Wirth (CO), Kent Conrad (ND), Howell Heflin (AL), John D. Rockefeller IV (WV).
*Minority* (Republicans): James A. McClure (ID), Mark O. Hatfield (OR), Pete V. Domenici (NM), Malcolm Wallop (WY), Frank H. Murkowski (AK), Don Nickles (OK), Conrad Burns (MT), Jake Garn (UT), Mitch McConnell (KY).

Subcommittee on Water and
Power
*Bill Bradley (NJ), Chairman*

## Committee on Environment and Public Works

Room SD-458 Dirksen Building
Washington, DC 20510
(202) 224-6176
*Quentin N. Burdick (ND),*
*Chairman*

Responsible for environmental policy,
environmental research, ocean dump-
ing, solid waste disposal, and recycling.
*Majority* (Democrats): Daniel P.
Moynihan (NY), George J. Mitchell
(ME), Max S. Baucus (MT), Frank R.
Lautenberg (NJ), John B. Breaux
(LA), Harry M. Reid (NV), Bob Gra-
ham (FL), Joseph I. Lieberman (CT).
*Minority* (Republicans): John H.
Chafee (RI), Alan K. Simpson (WY),
Steve Symms (ID), Dave Durenberger
(MN), John W. Warner (VA), James
M. Jeffords (VT), Gordon J. Hum-
phrey (NH).

Subcommittee on Energy
Regulation and Conservation
*Howard M. Metzenbaum (OH),*
*Chairman*

Subcommittee on Energy
Research and Development
*Wendell H. Ford (KY),*
*Chairman*

Subcommittee on Environmental
Protection
*Max S. Baucus (MT), Chairman*

Subcommittee on Mineral
Resources Development and
Production
*Jeff Bingaman (NM), Chairman*

Subcommittee on Nuclear
Regulation
*John B. Breaux (LA), Chairman*

Subcommittee on Public Lands,
National Parks, and Forests
*Dale L. Bumpers (AR),*
*Chairman*

Subcommittee on Superfund,
Ocean, and Water Protection
*Frank R. Lautenberg (NJ),*
*Chairman*

Subcommittee on Toxic
Substances, Environmental
Oversight, Research, and
Development
*Harry M. Reid (NV), Chairman*

Subcommittee on Water
Resources, Transportation,
and Infrastructure
*Daniel P. Moynihan (NY),*
*Chairman*

# STATE GOVERNMENT AND STATEWIDE ORGANIZATIONS

The groups listed in this section are
agencies and organizations that are
involved in a wide range of issues
and are usually not covered under
other headings. To locate state departments that have duties in one
specific area, please see the appropriate chapter: state agriculture departments, see *Land;* fish and
wildlife departments, see *Wildlife;*
forestry departments, see *Land;* energy departments, see *Energy;* education departments, see *Education.*

## Alabama

State Capitol
Montgomery, AL 36130
(205) 242-7100
*Guy Hunt, Governor*

## Alabama Conservancy

2717 7th Avenue South, Suite 201
Birmingham, AL 35233
(205) 322-3126
*Pete Conroy, President*

Non-government umbrella group encompassing the state's conservation organizations.

## Alabama Department of Conservation and Natural Resources

64 North Union Street
Montgomery, AL 36130
(205) 242-3486
*James D. Martin, Commissioner*

## Alabama Department of Environmental Management

1751 Congressman W. L.
Dickinson Drive
Montgomery, AL 36130
(205) 271-7700
*Cameron Vowell, Chairman*

## Alaska

P.O. Box A
Juneau, AK 99811
(907) 465-3500
*Steve Cowper, Governor*

## Alaska, Trustees for

725 Christensen Drive, Suite 4
Anchorage, AK 99501
(907) 276-4244
*Randall Weiner, Executive Director*

A public interest environmental law firm working to preserve Alaska's unique environmental values.

## Alaska Center for the Environment

700 H Street, Suite 4
Anchorage, AK 99501
(907) 274-3621

Non-government citizen-organizing center.

## Alaska Department of Environmental Conservation

P.O. Box 0
Juneau, AK 99811-1800
(907) 465-2600
*Dennis D. Kelso, Commissioner*

## Alaska Department of Natural Resources

400 Willoughby
Juneau, AK 99801
(907) 465-2400
*Lennie Boston Gorusch, Commissioner*

## Arizona

State House
Phoenix, AZ 85007
(602) 542-4331
*Fife Symington, Governor*

## Arizona Commission on the Arizona Environment

1645 West Jefferson, Suite 416
Phoenix, AZ 85007
(602) 542-2102
*Alicia Bristow, Executive Director*

Commission represents environmental and conservation groups in the government and the private sector.

## Arizona Department of Environmental Quality

2005 North Central Avenue
Phoenix, AZ 85004
(602) 257-2300
*Randolph Wood, Director*

## Arkansas

State Capitol
Little Rock, AR 72201
(501) 682-2345
*Bill Clinton, Governor*

## Arkansas Department of Pollution Control and Ecology

8001 National Drive, P.O. Box 9583
Little Rock, AR 72219
(501) 562-7444
*Randall Mathis, Director*

## California
State Capitol
Sacramento, CA 95814
(916) 445-2841
*Pete Wilson, Governor*

## California Coastal Commission
631 Howard Street
San Francisco, CA 94105
(415) 543-8555
*Peter Douglas, Executive Director*

South Coast District Office
245 West Broadway
Long Beach, CA 90802
(213) 590-5071

## California Department of Conservation
1416 Ninth Street, Room 1320
Sacramento, CA 95814
(916) 322-7683
*Randall M. Ward, Director*

## California Department of Consumer Affairs
107 South Broadway
Los Angeles, CA 90012
(800) 344-9940

## California Environmental Affairs Agency
P.O. Box 2815
Sacramento, CA 95812
(916) 322-5840
*Jananne Sharpless, Secretary, Environmental Affairs*

## California Public Interest Research Group (CALPIRG)
1147 South Robertson Boulevard
Los Angeles, CA 90035
(213) 278-9244
*Julie Duncan, Spokesperson*

Citizens' advocacy group that is also a member of a coalition of environmental organizations issuing Wastemaker awards to manufacturers who overpackage their products.

## California Resources Agency
1416 Ninth Street, Room 1311
Sacramento, CA 95814
(916) 445-5656
*Douglas P. Wheeler, Secretary*

The state's point man on environmental issues.

## Colorado
136 State Capitol
Denver, CO 80203
(303) 866-2471
*Roy Romer, Governor*

## Colorado Department of Natural Resources
1313 Sherman, Room 718
Denver, CO 80203
(303) 866-3311
*Hamlet J. Barry III, Executive Director*

## Colorado Environmental Coalition
777 Grant Street, #606
Denver, CO 80203-3518
(303) 837-8701
*Donald Thompson, President*

Citizens' coordinating council for environmental groups and individuals.

## Connecticut
State Capitol
Hartford, CT 06106
(203) 566-4840
*Lowell Weicker, Jr., Governor*

## Connecticut Council on Environmental Quality
Room 239, 165 Capitol Avenue
Hartford, CT 06106
(203) 566-3510
*Gregory A. Sharp, Chairman*

Prepares reports for the governor and addresses citizens' complaints.

**Connecticut Department of Environmental Protection**
State Office Building, 165 Capitol Avenue
Hartford, CT 06106
(203) 566-5599
*Leslie Carothers, Commissioner*

**Delaware**
Legislative Hall
Dover, DE 19901
(302) 736-4101
*Michael N. Castle, Governor*

**Delaware Department of Natural Resources and Environmental Control**
89 Kings Highway, P.O. Box 1401
Dover, DE 19903
(302) 736-4403
*Edwin H. Clark II, Secretary*

**Florida**
State Capitol
Tallahassee, FL 32399-0001
(904) 488-4441
*Lawton Chiles, Governor*

**Florida Department of Environmental Regulation**
2600 Blair Stone Road
Tallahassee, FL 32399-2400
(904) 488-4805
*Robert A. Mandell, Chairman*

**Florida Department of Fisheries and Aquaculture: Center for Wetlands**
University of Florida
Gainesville, FL 32611
(904) 392-1791
*Jerome V. Shiremen, Chairman*

**Florida Department of Natural Resources**
Marjory Stoneman Douglas Building
Tallahassee, FL 32399
(904) 488-1554
*Tom Gardner, Executive Director*

**Georgia**
State Capitol
Atlanta, GA 30334
(404) 656-1776
*Zell Miller, Governor*

**Georgia Conservancy**
781 Marietta Street, NW
Atlanta, GA 30318
(404) 872-8200
*E. Milton Bevington, Chairman*

Non-government organization that develops education programs and helps establish nature reserves.

**Georgia Department of Natural Resources**
Floyd Towers East, 205 Butler Street
Atlanta, GA 30334
(404) 656-3530
*J. Leonard Ledbetter, Commissioner*

**Georgia Environmental Council**
P.O. Box 2388
Decatur, GA 30031-2388
(404) 262-1967
*Bryan Ripley-Mager, President*

Non-government statewide umbrella of environmental organizations.

**Hawaii**
State Capitol
Honolulu, HI 96813
(808) 548-5420
*John D. Waihee, Governor*

**Hawaii Department of Land and Natural Resources**
P.O. Box 621
Honolulu, HI 96809
(808) 548-6550
*William W. Paty, Chairman*

**Hawaii Office of Environmental Quality Control**
465 South King Street, Room 104
Honolulu, HI 96813
(808) 948-7031
*Marvin T. Miura, Director*

## Idaho
State Capitol
Boise, ID 83720
(208) 334-2100
*Cecil D. Andrus, Governor*

## Idaho Environmental Council
P.O. Box 1708
Idaho Falls, ID 83403
(208) 336-4930
*Alan Hausrath, President*

Coordinates and stimulates financial resources and ideas among non-governmental environmental organizations.

## Illinois
State Capitol
Springfield, IL 62706
(217) 782-6830
*James Edgar, Governor*

## Illinois Department of Conservation
Lincoln Tower Plaza, 524 South Second Street
Springfield, IL 62701-1787
(217) 782-6302
*Mark Frech, Director*

## Illinois Environmental Protection Agency
2200 Churchill Road
Springfield, IL 62706
(217) 782-3397
*Bernard P. Killian, Director*

## Indiana
Room 206, Statehouse
Indianapolis, IN 46204
(317) 232-1048
*Evan Bayh, Governor*

## Indiana Department of Environmental Management
105 South Meridian Street
Indianapolis, IN 46225
(317) 232-8603
*Kathy Prosser, Commissioner*

## Indiana Department of Natural Resources
608 State Office Building
Indianapolis, IN 46204
(317) 232-4020
*James Lahey, Chairman*

## (Indiana) Hoosier Environmental Council
P.O. Box 1145
Indianapolis, IN 46206-1145
(317) 923-1800
*John Blair, President*

Non-government group which promotes more aggressive environmental regulations.

## Iowa
State Capitol
Des Moines, IA 50319
(515) 281-5211
*Terry Branstad, Governor*

## Iowa Department of Natural Resources
East Ninth and Grand Avenue, Wallace Building
Des Moines, IA 50319-0034
(515) 281-5145
*Larry J. Wilson, Director*

## Iowa Natural Heritage Foundation
Insurance Exchange Building, Suite 1005
505 Fifth Avenue
Des Moines, IA 50309
(515) 288-1846
*William J. Fultz, Chairman*

Serves as a catalyst and facilitator in the preservation and protection of Iowa's natural resources.

## Kansas
State House
Topeka, KS 66612
(913) 296-3232
*Joan Finney, Governor*

**Kansas Department of Health and Environment**
Landon State Office Building, Room 901
Topeka, KS 66612-1290
(913) 296-1522
*Stanley C. Grant, Secretary*

**Kentucky**
State Capitol
Frankfort, KY 40601
(502) 564-2611
*Wallace G. Wilkinson, Governor*

**Kentucky Environmental Quality Commission**
18 Reilly Road, Ash Annex
Frankfort, KY 40601
(502) 564-2150
*William Horace Brown, Chairperson*

**Kentucky Natural Resources and Environmental Protection Cabinet**
5th Floor, Capital Plaza Tower
Frankfort, KY 40601
(502) 564-3350
*Carl H. Bradley, Secretary*

**Louisiana**
State Capitol
Baton Rouge, LA 70804
(504) 342-7015
*Buddy Roemer, Governor*

**Louisiana Environmental Network**
555 St. Tammany Street
New Orleans, LA 70806
Non-government organization.

**Maine**
State House
Augusta, ME 04333
(207) 289-3531
*John R. McKernan, Jr., Governor*

**Maine Department of Conservation**
State House Station, #22
Augusta, ME 04333
(207) 289-2211
*C. Edwin Meadows, Jr., Commissioner*

**Maryland**
State House
Annapolis, MD 21404
(301) 974-3901
*William Donald Schaefer, Governor*

**Maryland Department of Natural Resources**
Tawes State Office Building
Annapolis, MD 21401
(301) 974-3041
*Torrey C. Brown, Secretary*

**Maryland Department of the Environment**
2500 Broening Highway
Baltimore, MD 21224
(301) 631-3000
*Martin W. Walsh, Jr., Secretary*

**Maryland Environmental Trust**
275 West Street, Suite 322
Annapolis, MD 21401
(301) 974-5350
*James B. Wilson, Chairman*

Non-government education, land-use planning, and conservation group.

**Massachusetts**
State House
Boston, MA 02133
(617) 727-3600
*William F. Weld, Governor*

**Massachusetts, Environmental Lobby of**
Three Joy Street
Boston, MA 02108
(617) 742-2553
*Ann Roosevelt, President*

Citizens' lobbying group.

## Massachusetts Executive Office of Environmental Affairs
Leverett Saltonstall Building
100 Cambridge Street
Boston, MA 02202
(617) 727-9800
*John DeVillars, Secretary*

## Michigan
State Capitol
Lansing, MI 48909
(517) 373-3400
*John Engler, Governor*

## Michigan Department of Natural Resources
Box 30028
Lansing, MI 48909
(517) 373-1220
*David F. Hales, Director*

## Michigan Nature Association
7981 Beard Road, Box 102
Avoca, MI 48006
(313) 324-2626
*Richard W. Holzman, President*

Non-government group which acquires and maintains nature sanctuaries.

## Minnesota
State Capitol
St. Paul, MN 55155
(612) 296-3391
*Arne Carlson, Governor*

## Minnesota Department of Natural Resources
500 Lafayette Road
St. Paul, MN 55155
(612) 296-6157
*Joseph N. Alexander, Commissioner*

## Minnesota Pollution Control Agency
520 Lafayette Road
St. Paul, MN 55155
(612) 296-6300
*Daniel D. Foley, Chairman of the Board*

## Mississippi
P.O. Box 139
Jackson, MS 39205
(601) 359-3100
*Ray Mabus, Governor*

## Mississippi Department of Environmental Quality

Bureau of Pollution Control
P.O. Box 10385
Jackson, MS 39289-0385
(601) 961-5171
*Charles H. Chisolm, Director*

Bureau of Land and Water Resources
Southport Mall, P.O. Box 10631
Jackson, MS 39289
(601) 961-5200
*Charles T. Branch, Bureau Director*

## Missouri
State Capitol
Jefferson City, MO 65101
(314) 751-3222
*John Ashcroft, Governor*

## Missouri Department of Conservation
P.O. Box 180
Jefferson City, MO 65102-0180
(314) 751-4115
*Jerry J. Presley, Director*

## Missouri Department of Natural Resources
P.O. Box 176
Jefferson City, MO 65102
(800) 334-6946
*G. Tracy Mehan III, Director*

## Montana
State Capitol
Helena, MT 59620
(406) 444-3111
*Stan Stephens, Governor*

**Montana Department of Natural Resources and Conservation**
1520 East Sixth Avenue
Helena, MT 59620-2301
(406) 444-6699
(406) 444-6721 FAX
*Karen L. Barclay, Director*

**Montana Environmental Quality Countil**
State Capitol
Helena, MT 59620
(406) 444-3742
*Bob Gilbert, Chairman*

**Nebraska**
State Capitol
Lincoln, NE 68509
(402) 471-2244
*Bill Nelson, Governor*

**Nebraska Department of Environmental Control**
301 Centennial Mall South
Lincoln, NE 68509
(402) 471-2186
*Dennis Grams, Director*

**Nebraska Natural Resources Commission**
301 Centennial Mall South,
P.O. Box 94876
Lincoln, NE 68509
(402) 471-2081
*Harold Kopf, Chairman*

**Nevada**
State Capitol
Carson City, NV 89710
(702) 885-5670
*Bob Miller, Governor*

**Nevada Department of Conservation and Natural Resources**
Capitol Complex, Nye Building,
201 South Fall Street
Carson City, NV 89710
(702) 885-4360
*Roland D. Westergard, Director*

**New Hampshire**
State House, Room 208
Concord, NH 03301
(603) 271-2121
*Judd Gregg, Governor*

**New Hampshire Council on Resources and Development**
c/o Office of State Planning
2½ Beacon Street
Concord, NH 03301
(603) 271-2155
*Jeffrey H. Taylor, Chairman*

**New Hampshire Department of Environmental Services**
Health and Human Services
Building, 6 Hazen Drive
Concord, NH 03301
(603) 271-2900
*Robert W. Varney, Commissioner*

**New Hampshire Department of Resources and Economic Development**
P.O. Box 856, 105 Loudon Road
Concord, NH 03301
(603) 271-2411
*Stephen K. Rice, Commissioner*

**New Hampshire State Conservation Committee: Department of Agriculture**
Caller Box 2042
Concord, NH 03302-2042
(603) 271-3576
*Robert M. Bodwell, Chairman*

**New Jersey**
State House, Office of the
Governor, CN-001
Trenton, NJ 08625
(609) 292-6000
*James Florio, Governor*

**New Jersey Conservation Foundation**
300 Mendham Road
Morristown, NJ 07960
(201) 539-7540
*I. Lloyd Gang, President*

Non-government group. Formerly the Great Swamp Committee.

## New Jersey Department of Environmental Protection
401 East State Street, CN 402
Trenton, NJ 08625-0402
(609) 292-2885
*Christopher J. Daggett, Commissioner*

## New Mexico
State Capitol
Santa Fe, NM 87503
(505) 827-3000
*Bruce King, Governor*

## New Mexico Energy, Minerals, and Natural Resources Department
525 Marquez Place
Santa Fe, NM 87503
(505) 827-5950
*Anita Lockwood, Cabinet Secretary*

## (New Mexico) Southwest Research and Information Center
P.O. Box 4524
Albuquerque, NM 87106
(505) 262-1862
*Don Hancock, Administrator*

Non-government group which publishes information on pollution in New Mexico.

## New York
State Capitol
Albany, NY 12224
(518) 474-8390
*Mario M. Cuomo, Governor*

## New York Department of Environmental Conservation
50 Wolf Road
Albany, NY 12233
(518) 457-3446
*Thomas C. Jorling, Commissioner*

## New York Environmental Protection Bureau
Department of Law
120 Broadway
New York, NY 10271
(212) 341-2451
*James Sevinsky, Assistant Attorney General in Charge*

## New York Office of Energy Conservation and Environmental Planning—New York State Department of Public Service
3 Empire State Plaza
Albany, NY 12223
(518) 474-1677
*Lawrence Dewitt, Director*

## North Carolina
State Capitol
Raleigh, NC 27603
(919) 733-4240
*James G. Martin, Jr., Governor*

## North Carolina Department of Environmental, Health, and Natural Resources
P.O. Box 27687
Raleigh, NC 27611
(919) 733-4984
*William W. Cobey, Jr., Secretary*

## North Dakota
600 East Boulevard, State Capitol, Ground Floor
Bismarck, ND 58505
(701) 224-2200
*George A. Sinner, Governor*

## Ohio
State House
Columbus, OH 43215
(614) 466-3555
*George V. Voinovich, Governor*

## Ohio Department of Natural Resources
Fountain Square
Columbus, OH 43224
(614) 265-6886
*Joseph J. Sommer, Director*

## Ohio Environmental Council
22 East Gay Street, Suite 300
Columbus, OH 43215-3113
(614) 224-4900
*Christine Carlson, President*

Non-government umbrella organization of environmental groups and individuals.

## Ohio Environmental Protection Agency
P.O. Box 1049, 1800 Watermark Drive
Columbus, OH 43266-0149
(614) 644-3020
*Richard L. Shank, Director*

## Oklahoma
State Capitol, Room 212
Oklahoma City, OK 73105
(405) 521-2345
*David Walters, Governor*

## Oklahoma Conservation Commission
2800 North Lincoln Boulevard, Suite 160
Oklahoma City, OK 73105
(405) 521-2384
*Hal Clark, Chairman*

## Oregon
State Capitol
Salem, OR 97310
(503) 378-3100
*Barbara Roberts, Governor*

## Oregon Department of Environmental Quality
811 SW Sixth Avenue
Portland, OR 97204
(503) 229-5300
*Fred Hansen, Director*

## Oregon Environmental Council
2637 SW Water Avenue
Portland, OR 97201
(503) 222-1963
*Allen Johnson, President*

Non-government coalition of environmental, conservation, health, and labor organizations to encourage legal and legislative action for the environment.

## Oregon Natural Resources Council
Yeon Building 1050, 522 Southwest Fifth Avenue
Portland, OR 97204
(503) 223-9001
*Tom Giesen, President*

Non-government umbrella organization including all aspects of conservation of natural resources.

## Pennsylvania
State Capitol
Harrisburg, PA 17120
(717) 787-2500
*Robert P. Casey, Governor*

## Pennsylvania Conservation Commission
Department of Environmental Resources
Executive House, Room 214, One Ararat Boulevard
Harrisburg, PA 17110
(717) 540-5080
*Arthur A. Davis, Chairman*

## Pennsylvania Department of Environmental Resources
Public Liaison Office
9th Floor, Fulton Building, P.O. Box 2063
Harrisburg, PA 17120
(717) 783-8303
*Arthur A. Davis, Secretary*

## Pennsylvania Environmental Council
1211 Chestnut Street, Suite 900
Philadelphia, PA 19107
(215) 563-0250
*Joseph M. Manko, President*

Statewide coalition to secure the passage and enforcement of environmental legislation and regulations.

## Puerto Rico
Conservation Trust of Puerto Rico
P.O. Box 4747
San Juan, Puerto Rico 00905
(809) 722-5834

Non-government organization.

## Puerto Rico Department of Natural Resources
P.O. Box 5887, Puerta de Tierra Station
San Juan, Puerto Rico 00906
(809) 724-8774
*Jose A. Labares Rivers, Secretary*

## Rhode Island
State House
Providence, RI 02903
(401) 277-2080
*Bruce Sundlun, Governor*

## Rhode Island Department of Environmental Management
9 Hayes Street
Providence, RI 02908
(401) 277-2774
*Robert L. Bendick, Jr., Director*

## South Carolina
State House
Columbia, SC 29211
(803) 734-9818
*Carroll A. Campbell, Jr., Governor*

## South Carolina Coastal Council
4130 Faber Place, Suite 300
Charleston, SC 29405
(803) 744-5838
*John C. Hayes III, Chairman*

Non-government group that protects the coastal environment and promotes responsible coastal development.

## South Carolina Department of Health and Environmental Control
J. Marion Sims Building, 2600 Bull Street
Columbia, SC 29201
(803) 734-4880
*Michael D. Jarrett, Commissioner*

## South Dakota
State Capitol
Pierre, SD 57501
(605) 773-3212
*George S. Mikelson, Governor*

## South Dakota Board of Minerals and Environment: Department of Water and Natural Resources
Joe Foss Building
Pierre, SD 57501
(605) 773-3151
*John J. Smith, Secretary of the Department*

Replaced Board of Environmental Protection and some of the activities of the South Dakota Conservation Committee.

## South Dakota Department of Water and Natural Resources
Joe Foss Office Building
Pierre, SD 57501
(605) 773-3151
*Floyd Mathew, Secretary*

## Tennessee
State Capitol
Nashville, TN 37219
(615) 741-2001
*Ned R. McWherter, Governor*

## Tennessee Department of Conservation
701 Broadway, Customs House
Nashville, TN 37219
(615) 742-6758
*Elbert T. Gill, Jr., Commissioner*

## Tennessee Environmental Council
1725 Church Street
Nashville, TN 37203
(615) 321-5075
*Joel Solomon, President*

Non-government group that coordinates the dissemination of information on environmental concerns for organizations and individuals.

## Texas
State Capitol
Austin, TX 78711
(512) 463-2000
*Ann Richards, Governor*

## Texas Committee on Natural Resources
5934 Royal Lane, Suite 223
Dallas, TX 75230-3803
(214) 368-1791
*Everett Harding, Chairman*

Non-government umbrella organization of volunteer task forces concerned with all aspects of conservation.

## Utah
210 State Capitol
Salt Lake City, UT 84114
(801) 538-1000
*Norman H. Bangerter, Governor*

## Utah Department of Natural Resources
1636 West North Temple
Salt Lake City, UT 84116-3154
(801) 538-7200
*Dee C. Hansen, Executive Director*
*Milo A. Barney, Associate Director*
*    for Resource Management*

## Vermont
Pavilion Office Building
Montpelier, VT 05602
(802) 828-3333
*Richard A. Snelling, Governor*

## Vermont Agency of Natural Resources
103 South Main Street
Waterbury, VT 05677
(802) 244-7347
*Jonathan Lash, Secretary*

## Virgin Islands (U.S.)
Government House
Charlotte Amalie
St. Thomas, U.S. Virgin Islands 00801
(809) 774-0001
*Alexander Farrelly, Governor*

## (Virgin Islands) Department of Planning and Natural Resources
Suite 231, Nisky Center
St. Thomas, U.S. Virgin Islands 00803
(809) 774-3320
*Alan Smith, Commissioner*

## Virginia
State Capitol
Richmond, VA 23219
(804) 786-2211
*L. Douglas Wilder, Governor*

## Virginia, Conservation Council of
P.O. Box 106
Richmond, VA 23201
(804) 384-1254
*Franklin F. Flint, President*

Non-government group that helps citizens and groups work with the legislative process.

## Virginia Council on the Environment
202 Ninth Street, Suite 900
Richmond, VA 23219
(804) 786-4500
*Keith J. Buttlemen, Administrator*

Implements the state environmental policy.

## Virginia Department of Conservation and Recreation
203 Governor Street, Suite 302
Richmond, VA 23219
(804) 786-2121
*B. C. Leynes, Jr., Director*
*Jerald F. Moore, Deputy Director*

## Virginia Marine Resources Commission
P.O. Box 756
Newport News, VA 23607
(804) 247-2200
*William A. Pruitt, Commissioner*

Regulates all commercial and sport-fishing, as well as the use of wetlands and primary sand dunes.

## Washington
State Capitol
Olympia, WA 98504
(206) 753-6780
*Booth Gardner, Governor*

## Washington Department of Ecology
Olympia, WA 98504
(206) 459-6000
*Andrea Beatty Riniker, Director*
*Phillip C. Johnson, Deputy Director, Programs*

Created in 1970 to promote programs and enforce the State Environmental Policy Act.

## Washington Department of Natural Resources
Public Lands Building
Olympia, WA 98504
(206) 753-5327
*Brian J. Boyle, Administrator*
*(Commissioner of Public Lands)*

Public Information Office
Public Lands Building
Olympia, WA 98504
Catalog available of environmental films.

## Washington Environmental Council
76 S. Main, P.O. Box 43445
Seattle, WA 98104
(206) 623-1483
*David Bricklin, President*

Non-government group which promotes citizen, legislative, and administrative action.

## West Virginia
State Capitol
Charleston, WV 25305
(304) 340-1600
*Gaston Caperton, Governor*

## West Virginia Department of Natural Resources
1900 Kanawha Boulevard, East
Charleston, WV 25305
(304) 348-2754
*J. Edward Hamrick, Director*

## West Virginia Highlands Conservancy
P.O. Box 306
Charleston, WV 25321
(304) 924-5802
*Cindy Rank, President*
*Donna Borders, Senior Vice-President*

Non-government group which promotes the conservation and wise use of West Virginia's natural and historic resources.

## Wisconsin
State Capitol
Madison, WI 53707
(608) 266-1212
*Tommy G. Thompson, Governor*

## Wisconsin Department of Natural Resources

P.O. Box 7921
Madison, WI 53707
(608) 266-2621
*C. D. Besadny, Secretary*

Bureau of Community
Assistance
P.O. Box 7921
Madison, WI 53707
(608) 266-5896
*Marjorie Devereaux, Director*

## Wyoming

State Capitol
Cheyenne, WY 82002
(307) 777-7434
*Mike Sullivan, Governor*

## Wyoming Environmental Quality Department

122 West 25th Street, 4th Floor
Cheyenne, WY 82002
(307) 777-7937
*Randolph Wood, Director*

## Wyoming Outdoor Council

201 Main Street
Lander, WY 82520
(307) 332-7031
*Kim Cannon, President*

Non-government group which monitors state and federal agencies.

# OTHER U.S. GOVERNMENT GROUPS AND INDIVIDUALS

## American College of Ecology

Route 1, Box 269
Clayton, IN 45118
(317) 539-4653

Lobbyists and legislative monitors for soil, water, and air pollution issues.

## American Committee for International Conservation

c/o Roger McManus,
   Secretary-Center for Marine
   Conservation
1725 DeSale Street, NW, Room 500
Washington, DC 20036
(202) 429-5609
*George Rabb, Chairman*

Exchanges information and coordinates activities of U.S. members of the International Union for Conservation of Nature.

## Americans for the Environment

1400 16th Street, NW
Washington, DC 20036
(202) 797-6665
*Betsy Loyless, Chairman of the Board*

Association of environmental groups to help citizens take part in the electoral process to improve the environment. Lobbyists.

## Boston Environmental Department

1 City Hall Square
Boston, MA 02129
(617) 725-3850
*Loraine Downey, Director*

## Bradley, Tom

200 N. Spring Street
Los Angeles, CA 90012

Mayor of Los Angeles, who has initiated the largest urban recycling program in the country.

## Earthcare Network

330 Pennsylvania Avenue, SE
Washington, DC 20003
(202) 547-1141
*Michael McCloskey, Chairman*

Environmental groups united for the purpose of providing mutual assistance in the conducting of environmental campaigns.

## Environment Action Fund (EAF)
P.O. Box 22421
Nashville, TN 37202
(615) 244-4994
*Jean Nelson, President*

Works for strong environmental legislative programs.

## Environmental Policy Institute
218 D Street, SE
Washington, DC 20003
(202) 544-2600
*Brent Blackwelder, Vice-President*

Wide-ranging lobbying organization.

## Environmental Safety
733 15th Street, NW, Suite 1120
Washington, DC 20005
(202) 628-0374
*Sherry Hiemstra, Executive Director*

Non-government advisory committee of former EPA officials and other environmental specialists who provide oversight on EPA activities.

## Green Committees of Correspondence
P.O. Box 30208
Kansas City, MO 64112
(816) 931-9366
*Dee Berry, Coordinator*

Seeks to develop a politically significant Green movement in the U.S.

## Green's Party
P.O. Box 127
Downers Grove, IL 60515
(312) 303-5996
*Randy Toler, Spokesperson*

Environmental politics specialists.

## Greenbelt Alliance
116 New Montgomery
San Francisco, CA 94104
(415) 543-4291
*Larry Orman, Executive Director*

Citizen group concerned with Bay Area environmental issues.

## Hayden, Tom
2141 State Capitol
Sacramento, CA 95814

State assemblyman and environmental activist.

## League of Conservation Voters
1150 Connecticut Avenue, NW, Suite 201
Washington, DC 20036
(202) 785-8683
*Jim Maddy, Executive Director*

Non-partisan group that lobbies for environmental issues and works to elect pro-environment candidates.

## League of Women Voters
1730 M Street, NW
Washington, DC 20036
(202) 429-1965
*Nancy M. Neumann, President*

Non-partisan political organization that lobbies for environmental causes.

## National Association of Conservation Districts
509 Capitol Court, NE
Washington, DC 20002
(202) 547-6223
*Robert Wetherbee, President*

Links 3,000 local districts and 54 state and territorial associations.

North Central Region
1053 Main Street
Stevens Point, WI 54481
(715) 341-1022
*William J. Horvath, Regional Representative*

Northeastern Region
804 South State Street,
P.O. Box 260
Dover, DE 19903-0260
(302) 734-7377
*Lynn Sprague, Regional Representative*

Pacific Region
831 Lancaster Drive, NE,
Suite 207
Salem, OR 97301
(503) 363-0912
*Robert C. Baum, Regional
Representative*

Southern Region
11318 North May,
Suite D
Oklahoma City, OK 73120
(405) 755-2288
*Robert W. Toole, Regional
Representative*

Western Region
9150 West Jewell Avenue,
Suite 113
Lakewood, CO 80226
(303) 988-1810
*Robert E. Raschke, Regional
Representative*

**Public Interest Research Group
(PIRG)**
215 Pennsylvania Avenue, SE
Washington, DC 20003
(202) 546-9707
*Ralph Nader, Founder*
*Gene Karpinski, Executive Director*

Lobbies for reform on consumer, environmental, energy, and governmental issues.

**Resource Policy Institute**
P.O. Box 39185
Washington, DC 20016
(202) 686-8875
*Arthur Purcell, Director*

Lobbies for science compatible with the environment, concerned with water, minerals, resource conservation, energy and technology policy, and waste issues.

**U.S. Man and the Biosphere
Program**
U.S. MAB Secretariat,
OES/ENR/MAB, Department
of State
Washington, DC 20520
(202) 632-2786
*Thomas E. Lovejoy, Chairman*

Government research projects on High-Latitude Ecosystems, Human Dominated Systems, Marine and Coastal Ecosystems, Temperate Ecosystems, and Tropical Ecosystems.

**Urban Environment Conference**
815 16th Street, NW
Washington, DC 20006
(202) 638-6429
*Franklin Wallick, Chairman*

Non-government organization made up of national labor, minority, and environmental organizations who lobby for strong environmental and occupational health laws.

# MULTI-NATIONAL ORGANIZATIONS

**Canada-United States
Environmental Council**

Canada
c/o Canadian Nature Federation
75 Albert Street
Ottawa, Ontario K2P 6G1,
Canada
(613) 238-6154
*Paul Griss, Co-Chairman*
Non-government organization of Canadian and U.S. environmental groups.

United States
c/o Defenders of Wildlife
1244 19th Street, NW
Washington, DC 20036
(202) 659-9510
*James G. Deane, Co-Chairman*

## International Joint Commission

United States
2001 S Street, NW, Second
  Floor
Washington, DC 20440
(202) 673-6222
*Gordon K. Durnil, U.S.*
  *Chairman*
Created by the U.S. and Great
  Britain in 1909 to prevent and
  mediate water disputes on the
  boundaries of Canada and the
  U.S.

Canada
100 Metcalfe Street
Ottawa, Ontario K1P 5M1
  Canada
(613) 995-2984
*E. Davie Fulton, Canadian*
  *Chairman*

## Nordic Council for Ecology

c/o Pehr H. Enckell, Ecology
  Building
University of Lund
S-223 62 Lund, Sweden
14 81 88

Promotes education and research in
Nordic countries.

## South Asia Cooperative Environment Programme

P.O. Box 1070
Colombo 5, Sri Lanka
589369

Promotes cooperation among countries
of South Asia in environmental protection.

## World Environment Center

4198 Park Avenue, Suite 1403
New York, NY 10016
(212) 683-4700
*Antony G. Marcil, President*

Serves as bridge between industry and
governments.

# FOREIGN ORGANIZATIONS

## Afghanistan

### Ministry of Agriculture and Land Reform

Department of Environmental
  Protection
Kabul, Afghanistan
41151

## Algeria

### Agence Nationale pour la Protection de l'Environnement

B.P. 154
El-Annaser, Algeria
771414

National Agency for the Protection of
the Environment.

## Antarctica

### Antarctica and Southern Ocean Coalition

201 D Street, NW
Washington, DC 20003
(202) 544-2600
*Jim Barnes, Co-Director*

Works to protect the Antarctic region,
monitors governments and holds them
accountable.

### Antarctica Project

218 D Street, SE
Washington, DC 20003
(202) 544-2600
*Jim Barnes, Executive Director*

Purpose is to preserve the Antarctic
region and its wildlife by securing
moratorium on minerals activities,
regulating fishing, establishing whale
sanctuaries, etc.

**Central Administration of the French Southern and Antarctic Lands**
34 rue des Renaudes
F-75017 Paris, France

**Commission for Conservation of Antarctic Marine Living Resources**
25 Old Wharf
Hobart, Tasmania 7000, Australia
310366
23 2714 FAX

**Instituto Antártico Chileno**
Avenida Luis Thayer Ojeda 814
Santiago, Chile
231-0105
Chilean Antarctic Institute.

## Argentina

**Secretaría General de la Presidencia, Subsecretaría de Política Ambiental**
Corrientes 1302, 1er piso
1043 Buenos Aires, Argentina
40-6721

General Secretariat of the Presidency, Subsecretariat of Environmental Policy.

## Australia

**Department of Arts, Sport, the Environment, Tourism and Territories**
GPO Box 1252
Canberra, ACT 2601 Australia
62 46-7211
62 48-0334 FAX
*W. J. Harris, Associate Secretary*

**National Parks and Wildlife Service**
G.P.O. Box 636
Canberra, ACT 2601 Australia
46-6008

## Austria

**Ministry for Environment, Youth, and Family**
Biberstrasse 11
A-1010 Vienna, Austria
222 52 35 21
*Marilies Fleming, Minister*

## Bahamas

**Bahamas National Trust**
P.O. Box N-4105
Nassau, The Bahamas

Non-government group that works to preserve natural and historic areas.

**Ministry of Agriculture, Trade and Industry**
P.O. Box N-3028
Nassau, The Bahamas
(809) 323-1777
*Perry G. Christie, Minister*
*Basil O'Brien, Permanent Secretary*

## Bangladesh

**Department of Environment Pollution Control**
6/11/F Lalmatia Housing Estate, Satmasjid Road
Dhaka, Bangladesh
31 57 77

## Belgium

**Department of State for Environment and Social Emancipation**
Wetstraat 56
B-1040 Brussels, Belgium
02/2304985
*Mrs. Miet Smet, Secretary*

## Belize

### Belize Center for Environmental Studies
P.O. Box 666
Belize City, Belize

Non-government public education forum.

## Bermuda

### Ministry of the Environment
Government Administration Building
30 Parliament Street
Hamilton HM 12, Bermuda

## Bolivia

### Comisión del Medio Ambiente y Recursos Naturales
P.O. Box 8561
La Paz, Bolivia
37-4265

Commission on the Environment and Natural Resources.

## Botswana

### Department of Wildlife and National Parks
P.O. Box 131, Private Bag 004
Gaborone, Botswana
51461
*K. Ngwamotsouko, Permanent Secretary, Ministry of Commerce and Industry*

## Brazil

### Ministério de Agricultura, Instituto Brasileiro de Desenvolvimento Florestal
Sector das Areas Isoladas Norte, Avenida L4 Norte
Brasilia DF, Brazil
061-1711

Ministry of Agriculture, Brazilian Institute for Forestry Development.

### Ministério de Habitaçãtiao, Urbanismo e Meio Ambiente Secretaria Especial do Meio Ambiente
Avenida W3 Norte—Q510, Edifício Cidade de Cabo Frio
Brasilia DF 70.750 Brazil
274-6515

Ministry of Housing, Urbanism, and Environment—Special Secretariat of the Environment

### President Fernando Collor de Mello
c/o Brazilian Embassy
3006 Massachusetts Avenue, NW
Washington, DC 20008
(202) 797-0200

## Bulgaria

### Committee For Environment Protection
Ul. Triaditza 2
Sofia 1000, Bulgaria
87-61-51
52-16-34 FAX
*Alexander Alexandor, Minister*

## Burma

### Ministry of Health and Information
Rangoon, Burma
*U Tan Wai, Minister*

Government department in charge of conservation and environmental affairs.

## Cambodia

**Ministère de l'Economie et des Finances, Département de l'Environment**
c/o Permanent Mission of Democratic Kampuchea to ESCAP
United Nations Building
Rajadamnern Avenue, Bangkok 2, Thailand

Ministry of the Economy and Finance, Department of the Environment.

## Cameroon

**Service de la Faune et des Parcs Nationaux**
Yaounde, Cameroon

Wildlife and National Park Service.

## Canada

**British Columbia—Ministry of the Environment**
Parliament Buildings
Victoria, British Columbia V8V 1X5 Canada

**Canadian Environmental Network**
P.O. Box 1289, Station B
Ottawa, Ontario K1P 5R3 Canada
(613) 563-2078

Information-sharing network of more than 1,000 environmental groups across Canada.

**Department of the Environment**
Ottawa, Ontario K1A 1G2 Canada
(819) 997-2800

**Environment Canada**
Ottawa, Ontario K1A 1G2 Canada
(819) 997-2800

Government agency responsible for the Forestry Service, Wildlife Service, national parks and most other environmental concerns.

**Ontario Ministry of the Environment**
10 Wellington Street
Ottawa, Ontario K14 OH3 Canada
(819) 997-1441
*Lucien Bouchard, Minister*

## Chad

**Ministere du Tourisme et de l'Environment**
B.P. 447
N'Djamena, Chad
2121

Ministry of Tourism and the Environment.

## Chile

**Corporación Nacional Forestal y de Protección de Recursos Naturales Renovables**
Avenida Bulnes 285, Oficina 501
Santiago, Chile
696-0783
*Armando Alvarez, Minister of Natural Resources*

National Corporation for Forestry and Protection of Natural Resources.

## China

**Chinese Society for Environmental Sciences: Division of Nature Protection**
c/o Environment Protection Bureau
Ministry of Urban Construction and Rural Construction and Environmental Protection
Bainwanzhuang, Beijing, China

**Research Center for Eco-Environmental Sciences**
P.O. Box 934
Beijing, China
28-5176

Non-government organization.

## China (Taiwan)

**Ministry of the Interior, Section 4**
107 Roosevelt Road
Taipei, Taiwan, Republic of China
*Chuang-huan Chiu, Minister*

## Colombia

**Instituto Nacional de los Recursos Naturales Renovables y del Ambiente**
Apartado Aerero 13458
Bogota, D.E., Colombia
285-4417

National Institute of Natural Resources and the Environment.

## Congo

**Ministère du Tourisme, Loisirs et de l'Environnement**
B.P. 958
Brazzaville, Congo
81-30-46

Ministry of Tourism, Recreation and the Environment.

## Costa Rica

**Departamento de Conservación y Reforestación**
Apartado Postal 10032
1000 San Jose, Costa Rica

Department of Conservation and Reforestation.

## Cuba

**Academia de Ciencias de la República de Cuba**
Capitolio Nacional
Havana 2, Cuba
68914

Academy of Sciences of the Republic of Cuba.

## Czechoslovakia

**Czech Ministry of Interior and Environment**
Mala Strana, Waldstejnske
Namesti 1
CS-118 Prague, Czechoslovakia

**Slovak Ministry of Interior and Environment**
Mostova 6
CS-811 02 Bratislava,
Czechoslovakia

## Denmark

**Miljoestyrelsen**
Strandgade 29, DK-1401
Copenhagen K, Denmark
57 83 10

Agency of Environmental Protection.

## Ecuador

**Consejo Nacional de Desarrollo, Programa de Recursos Naturales y Medio Ambiente**
Quito, Ecuador

National Council of Development, Program of Natural Resources and Environment.

## Egypt

**Environmental Affairs Agency**
Cabinet of Ministers, Kasr Al-Aini Street
Cairo, Egypt

## El Salvador

**Comité Nacional de Protección del Medio Ambiente**
c/o Ministerio de Planificación
San Salvador, El Salvador

National Committee of Environmental Protection.

## Ethiopia

**Ethiopian Wildlife
Conservation Organization**
P.O. Box 386
Addis Ababa, Ethiopia

Non-government group.

## Fiji

**Ministry of Agriculture and
Fisheries**
P.O. Box 358
Suva, Fiji
211515

## Finland

**Ministry of the Environment**
Ratakatu 3, P.O. Box 399
SF-00531 Helsinki, Finland
19911
1991 499 FAX
*Kaj Barlund, Minister*

## France

**Fédération Française des
Sociétés de Protection de la
Nature**
57 rue Cuvier
F-75231 Paris, France

Non-government French Federation of
Nature Protection Societies.

**Ministère de l'Environment et
du Cadre de Vie**
14 boulevard du General-Leclerc
F-92524 Neuilly-sur-Seine, France
1-47-58-12-12
1-47-45-04-74 FAX
*Brice Lalonde, Secretary of State to
the Prime Minister, responsible
for the environment.*

Ministry of the Environment.

## French Polynesia

**Union Polynésienne pour la
Sauvegarde de la Nature**
Te Rauatiati a tau a Hiti Noa Tu
B.P. 1602
Papeete, Tahiti

Polynesian Union for Protection of Nature (May Nature Be Conserved Forever).

## Germany

**Bund für Umwelt und
Naturschutz Deutschlands**
Postfach 12 05 36
D-5300 Bonn, Germany

Non-government Union for Environmental and Nature Protection.

**Die Grünen**
Bundeshaus
HT 718, 5300 Bonn 1, Germany
0228-167918
0228-169206

The first environmental political party.

**Federal Ministry for the
Environment, Nature
Conservation, and Nuclear
Safety**
Adenauerallee 4
5300 Bonn, Germany
681-4164
*Klaus Topfer, Minister*

**Schutzgemeinschaft Deutches
Wild**
Meckheimerallee 79
D-5300 Bonn 1, Germany
658 462

Non-government German Association
for the Protection of Forests and
Woodlands.

## Ghana

**Environmental Protection Council**
P.O. Box M326, Ministries' Post Office
Accra, Ghana
66 46 97

## Greece

**Penellhnio Kentro Oikologikon Ereynon**
7 Sufliou Street, Ampelokopi
GR-115 27 Athens, Greece

Non-government Panhellenic Center of Environment Studies.

## Greenland

**Greenland Home Rule Government—Secretariat for Fisheries and Industry**
P.O. Box 1015
3900 Nuuk, Greenland
299-2-30-00

## Guatemala

**Comisión Nacional del Medio Ambiente**
9a Avenida entre 14 y 15, Zona 1, Oficina 10
Edificio Antigua Corte Suprema
Guatemala City, Guatemala
2 2-1816
*Jorge Cabrera Hidalgo, Coordinator*

National Environment Commission.

## Guinea

**Direction de l'Environment**
B.P. 3118
Conakry, Guinea
46 10 12

Environment Directorate.

## Guyana

**Ministry of Health and Public Welfare**
Homestretch Avenue, D'Urban Park
Georgetown, Guyana
2 65 861
*Noel Blackman, Minister*

## Honduras

**Ecology Association**
P.O. Box T-250
Tegucigalpa, D.C., Honduras
32-90-18

## Hungary

**Ministry for Environment and Water Management**
Fo-u. 46-50, Pf 351
H-1011 Budapest, Hungary
166-245

Includes the Institute for Environmental Protection.

## Iceland

**Nature Conservation Council**
Hverfisgotu 26
ISL-101 Reykjavik, Iceland
22520

## India

**Ministry of Environment and Forests**
B Block, Paryavaran Bhavan, CGO Complex, Lodi Road
New Delhi 110003 India
*Maneka Gandhi, Minister*

## Indonesia

**Directorate of Nature Conservation**
Department of Forestry, Jlan lr. H., Juanda No. 9
Bogor—West Java, Indonesia
*Effendi Ahmad Sukmaredja, Director*

**Ministry of State for Population and Environment**
Jalan Merdeka Barat no. 15, 3rd Floor
Jakarta, Indonesia
371-295

## Iran

**Department of the Environment**
Ostad Nejat-Ollahi, Avenue no. 187
P.O. Box 4565-15875
Teheran, Iran
891261

## Iraq

**Higher Council for Environmental Protection and Improvement**
c/o Ministry of Health
P.O. Box 423
Baghdad, Iraq

## Ireland

**Department of the Environment**
Custom House
Dublin 2, Ireland

## Israel

**Environmental Protection Service**
Ministry of the Interior
P.O. Box 6158
Jerusalem 91061 Israel
2 660151
*Uri Marinov, Director*

**Society of the Protection of Nature in Israel**
4 Hasfela Street
Tel Aviv 66183, Israel
375063
Non-government group.

## Italy

**Ministry of the Environment**
Piazza Venezia 11
I-00187 Rome, Italy
67 97 124

## Jamaica

**Natural Resources Conservation Department**
P.O. Box 305, 53½ Molynes Road
Kingston 10, Jamaica
923-5155
*Ted Aldridge, Minister*

## Japan

**Environment Agency of Japan**
1-2-2 Kasumigaseki
Chiyoda-ku Tokyo 100 Japan
580-4982
*Ishimatsu Kitagawa, Director General*

## Jordan

**Ministry of Municipal and Rural Affairs and the Environment**
Department of the Environment
P.O. Box 35206
Amman, Jordan
*Yusuf Hamdan Al Jaber, Minister*

**Royal Society for the Conservation of Wildlife**
Jebel Amman, Fifth Circle
Amman, Jordan
Non-government society.

## Kenya

**Ministry of Tourism and Wildlife**
Utalii House, Box 30027
Nairobi, Kenya
891601

**Ministry of Environment and Natural Resources**
P.O. Box 67839
Nairobi, Kenya
332383

**Kenya Wildlife Services**
P.O. Box 40241
Nairobi, Kenya

## Korea (North)

**Natural Conservation Union of the Democratic People's Republic of Korea**
No. 220-93-7-24 Dongsong Street
Central District, Pyongyang, Korea

## Korea, Republic of (South Korea)

**Environment Administration**
17-16 Sincheon-dong, Songpa-gu
Seoul 134-240, Korea

## Lesotho

**National Environment Secretariat**
Prime Minister's Office
P.O. Box 527
Maseru, Lesotho
23861

## Liberia

**Forestry Development Authority**
P.O. Box 3010
Sinkor, Monrovia, Liberia

## Libya

**National Committee for the Protection of the Environment**
Tripoli, Libya

## Liechtenstein

**Ministry of Agriculture, Forestry, and Environnment**
FL-9490 Vadus, Liechtenstein

## Luxembourg

**Ministère de l'Environnement et des Eaux et Forêts**
5a rue de Prague
Luxembourg-Ville, Luxembourg
44 15 08
40 04 10 FAX
*M. Alex Bodry, Minister of Environment*
Ministry of the Environment, Water, and Forests.

## Madagascar

**Association pour la Sauvegarde de l'Environnement**
B.P. 412
Antisiranana, Madagascar
Association for Environmental Protection.

## Malawi

**Office of the President, Environmental Department**
Private Bag 388
Lilongwe 3, Malawi
722780

## Malaysia

**Ministry of Science, Technology, and the Environment**
Wisma Sime Darby, Jalan Raja Laut
50662 Kuala Lumpur, Malaysia
60-3 293-8955
*Datuk Amar Stephen Yong, Minister*

## Malta

**Department of Health, Environment Division**
Valletta, Malta
22 40 71

## Mauritania

**Ministry of Rural Development**
B.P. 170
Nouakchott, Mauritania
51836
*Oumar Ba, Minister*

## Mauritius

**Ministry of Agriculture, Fisheries, and Natural Resources**
Reduit, Mauritius
4-1091

## Mexico

**Pacto de Grupos Ecologistas**
Amores 1814, Colonia del Valle
03100 Mexico City, D.F., Mexico
534-1023

Non-government Pact of Environmental Groups.

**Secretaría de Desarrollo Urbano y Ecología**
Avenida Constituyentes 947,
Edificio B, PB
Mexico City, D.F., Mexico
271-0355
*Sergio Reyes Lujan, Secretary of Ecology*

Secretariat of Urban Development and Ecology.

## Mongolia

**State Committee on Science and Technology**
Ulan Bator, Mongolia
29957

## Morocco

**Minister of Housing, Urban Development, and Environment**
Rabat 60263 Morocco
*Abderrahmane Boufettass, Minister*

## Namibia

**Department of Agriculture and Nature Conservation**
Private Bag 13306
Windhoek 9000 Namibia

## Nepal

**National Commission for Conservation of Natural Resources**
Babarmahal, Kathmandu, Nepal

## Netherlands

**Ministerie van Landbouw en Visserij**
Postbus 204011
NL-200 EK, The Hague, The Netherlands
31-70-793266
*J. B. Pieters, Director for Nature, Environment and Wildlife Management*

Ministry of Agriculture and Fisheries.

## New Zealand

**Department of Conservation**
P.O. Box 10-420
Wellington, New Zealand
64-4 710-726
*David McDowell, Director-General*

**Ministry for the Environment**
P.O. Box 10362
Wellington, New Zealand
734-090
710-195 FAX

## Nicaragua

**Instituto Nicaraguense de Recursos Naturales y del Ambiente**
Apartado 512312
Managua, Nicaragua
(505 2) 3-1110

Nicaraguan Institute of Natural Resources and the Environment.

## Niger

**Ministère de l'Hydraulique et de l'Environnement**
B.P. 578
Niamey, Niger
73 33 329
*Brigi Rafini, Minister*

Ministry of Water Resources and the Environment.

## Nigeria

**Federal Ministry of Works and Housing, Environmental Planning and Protection Division**
P.M.B. 12698, Lagos, Nigeria
682625

**Nigerian Conservation Foundation**
Mainland Hotel, 1st Floor,
P.O. Box 467
Lagos, Nigeria
*S.L. Edu, Chief, President and Trustee*

Non-government group.

## Norway

**Royal Ministry of Environment**
Myntgaten 2, Postboks 8013-Dep.
N-0030, Oslo 1, Norway
34 57 15
11 60 08 FAX

## Oman

**Council for Conservation of Environment and Water Resources**
P.O. Box 5575
Ruwi, Oman
704006
*Sayyid Shabib bin Taimur bin Faisal Al-Said, Minister*

## Pakistan

**National Council for Conservation of Wildlife in Pakistan**
485 Street 84
G-6/4 Islamabad, Pakistan

## Panama

**Fundación de Parques Nacionales y Medio Ambiente**
Apartado 6-6623
El Dorado, Panama
25-3676

Non-government Foundation for National Parks and the Environment.

**Ministerio de Planificación y Política Económica, Comisión Nacional del Medio Ambiente**
Apartado 2694
Zona 3, Panama City, Panama
(507) 32-6055

Ministry of Planning and Economic Policy, National Environment Commission.

## Papua New Guinea

**Department of Environment and Conservation**
Central Government Buildings, Waigani
P.O. Box 6601
Boroko, Papua New Guinea
22327

## Paraguay

**Ministerio de Agricultura y Ganadería**
Presidente Franco y 14 de Mayo
Asunción, Paraguay
*Hernando Bertoni, Minister of Agriculture*

Ministry of Agriculture and Livestock.

## Peru

**Centro de Investigación y Promoción Amazonica**
Avenida Ricardo Palma 666-D
Lima 18, Peru
46-4823

Center for Amazonian Research and Promotion.

**Ministerio de Agricultura, Dirección General Forestal y de Fauna**
Natalio Sanchez 220, 3er piso,
Lima 11, Peru
23-3978
*Juan M. Coronado, Minister of Agriculture*
*Wiley Harm Sparza, Minister of Fisheries*

Ministry of Agriculture, General Directorate of Forestry and Wildlife.

## Philippines

**Department of the Environment and Natural Resources**
Quezon Avenue, Diliman
Quezon City 1100, Philippines
2 976-626
*Fulgencio Factoran, Jr., Secretary*

## Poland

**Ministerstwo Ochrony Srodowiska i Zasobow Naturalnych**
ul. Wawelska 52–54
PL-02-067 Warsaw, Poland
299254
*Bronislaw Kaminski, Minister*

Ministry of Environmental Protection and Natural Resources.

**Polish Green Party**
P.O. Box 783
PL-30-960 Krakow, Poland

Non-government political group focusing on environmental issues.

**Polski Kluby Ekologiczne**
Rynek Glowny 27
PL-31-010 Krakow, Poland

Non-government Polish Ecological Clubs, affiliated with Friends of the Earth.

## Portugal

**Secretaria de Estado do Ambiente e dos Recursos Naturais**
Rua do Seculo, 51-20
P-1200 Lisboa, Portugal
*Carlos Alberto Martins Pimenta, Secretaria*

Secretary of State for Environment and Natural Resources.

## Romania

**National Council for Environmental Protection**
Piata Victoriei 1
Bucharest, Romania
14 34 00

## Rwanda

**Office Rwandais du Tourisme et des Parcs Nationaux**
Présidence de la République, B.P. 905
Kigali, Rwanda
*Laurent Habiyaremye, Director*

Rwandan Office of Tourism and National Parks.

## Saudi Arabia

**Meteorology and Environmental Protection Administration**
Makarona Street, P.O. Box 1358
Jeddah, Saudi Arabia
665-0084
*Abdul Bin al-Qeen, Chairman*
*Rumaih Mansour al-Rumaih, Director-General*

## Senegal

**Association Sénégalaise des Amis de la Nature**
B.P. 1801
Dakar, Senegal
222573

Non-government Senegalese Association of Friends of Nature.

**Direction de l'Environnement**
Dakar, Senegal

Environment Directorate.

## Seychelles

**Ministry of National Development**
Independence House, P.O. Box 199
Victoria, Mahe, Seychelles
22881

## Sierra Leone

**Chief Conservator of Forests, Ministry of Agriculture and Forests**
Youki Building
Brookfields, Freetown, Sierra Leone
24821

## Singapore

**Ministry of Environment**
Princess House, Alexandra Road
Singapore 0315
635111

## Somalia

**National Range Agency**
P.O. Box
Mogadishu, Somalia

## South Africa

**Department of Water Affairs, Forestry, and Environment Conservation**
Private Bag X313
Pretoria 0001 South Africa

## Spain

**Instituto Nacional para la Conservación de la Naturaleza**
Gran Via de San Francisco
85 Madrid 28005 7 Spain
266-8200
*Santiago Marraco Solana, Director*
National Institute for the Conservation of Nature.

## Sri Lanka

**Central Environmental Authority**
Maligawatte, New Town
Colombo 10, Sri Lanka
549455

## Sudan

**Ministry of Agriculture**
Wildlife Conservation and National Parks
P.O. Box 336
Khartoum, Sudan
76486

## Sweden

**National Swedish Environment Protection Board**
Box 1302
S-171 25 Solna 1, Sweden
799 10 00
29 23 82 FAX
*Valfrid Paulsson, Director-General*

## Switzerland

**Federal Office for Environmental Protection**
CH-3003
Bern, Switzerland
61 93-23
619981 FAX

## Syria

**Ministry of State for Environmental Affairs**
Government House
Damascus, Syria
11 22 66
*Kamel Elbaba, Minister*

## Tanzania

**Chama cha Kuendeleza Mazingira na Viumbe Hai Tanzania**
P.O. Box 1309
Dar es Salaam, Tanzania
Non-government Tanzanian Environmental Society.

**Ministry of Lands, Housing, and Urban Development**
National Environment Management Council
P.O. Box 20671
Dar Es Salaam, Tanzania
21241

## Thailand

**National Environment Board**
Soi Prachasumpun 4, Rama VI Road
Bangkok 10400, Thailand
278-5476

**Royal Forestry Department**
Phaholyothin Road
Bangkok 10900, Thailand
579-1587
*Chumni Boonyophas, Director-General*

## Togo

**Ministère de l'Environnement
et du Tourisme**
B.P. 355
Lome, Togo
Ministry of Environment and Tourism.

## Tunisia

**Ministère de l'Agriculture et de
l'Environnement**
30 rue Alain Savary
Tunis, Tunisia

Ministry of Agriculture and the Environment.

## Turkey

**Directorate General for
Environmental Affairs**
Atatturk Bulvari 143
Bakanliklar-Ankara, Turkey
*Musaffer Evirgen, Director-General
Kazim Ceylan, Assistant
Director-General*

## Uganda

**Ministry of Environment
Protection**
P.O. Box 8147
Kampala, Uganda

## Union of Soviet
## Socialist Republics

**Ministry of Atomic Energy**
Staromonetny per.26
Moscow, USSR

**State Committee of the USSR
for Nature Protection**
11 Nezhdanova Street
Moscow, USSR
*Nikolai N. Vorontson, Chairman*

**Vserossiiskoe Obshchestvo
Okhrany Priody**
Kuibyshevskii prospekt 3
Moscow L-12, USSR
Non-government All-Russian Society
for Nature Conservation, focuses
largely on environmental education.

## Union of Soviet Socialist
## Republics (Estonia)

**Eesti Roheline Liikkumine**
P.O. Box 3207
200080 Tallinn, Estonian SSR
Non-government Estonian Green
Movement.

## Union of Soviet Socialist
## Republics (Latvia)

**VAK Environment Protection
Club**
c/o Leipajan Pedagogical
Institute, Klaipedas 68-7
229700 Liepaja, Latvian SSR
The non-governmental Greens of
Latvia.

## Union of Soviet Socialist
## Republics (Ukraine)

**Zelenyi Swit**
ul. Belgorodskaya 8, Apt. 55
252137 Kiev, Ukrainian SSR
Non-government environmental protection agency whose name means
"Green World."

## United Arab Emirates

**Higher Environmental
Committee**
c/o Ministry of Health
P.O. Box 1853
Dubai, United Arab Emirates
23 30 21

## United Kingdom

**Department of Energy**
Thames House South
Millbank, London SW1P 4QJ
England
211 3000

**Ecology Party**
36-38 Clapham Road
London SW9 OJQ England
735 2485

Non-government political group, Green Party of the U.K.

**Green Alliance**
60 Chandos Place
London WC2N 4HG England

Works to build a constituency in each political party in Great Britain to promote the environment.

**Minister of State, Department of the Environment**
2 Marsham Street
London SW1P 3EB England
01-212-3434
*John Gummer, Minister for Environment*
*Michael Howar, Secretary of State for the Environment*

**Royal Commission on Environmental Pollution**
Church House, Great Smith Street
London SW1P 3BL England
2128620

## Uruguay

**Ministerio de Ganadería, Agricultura y Pesca**
Dirección General de Recursos Naturales Renovables
Cerrito 322, 2 piso
Montevideo, Uruguay
95 98 78

Ministry of Livestock, Agriculture, and Fisheries, General Directorate of Renewable Natural Resources.

## Venezuela

**Ministerio de Ambiente y de los Recursos Naturales Renovables**
Torre Norte, Centro Simón Bolívar
Caracas, Venezuela
483-3164

Ministry of the Environment and Renewable Natural Resources.

## Vietnam

**Comité d'État des Sciences et Techniques**
39 rue Tran Hung Dao
Hanoi, Vietnam
52731

State Committee for Science and Technology.

## Virgin Islands (British)

**Ministry of Natural Resources and Labour**
Road Town, Tortola, British Virgin Islands
43701

## Yugoslavia

**Savez za Zastitu Covekoves Sredine**
2 boulevar Lenjina
YU-11000 Belgrade, Yugoslavia

Federal Agency for the Human Environment.

## Zaire

**Département de l'Environnement, Conservation de la Nature et Tourisme**
B.P. 12348
Kinshasa 1, Zaire
31252
*Pendje Demodeto Yako, Commissioner*

Department of the Environment, Nature Conservation, and Tourism.

## Zambia

**Department of National Parks and Wildlife Service**
Private Bag 1
Chilanga, Zambia
01/278366
*Patrick M. Chipungu, Director*
*Ackim Mwenya N., Deputy Director*

## Zimbabwe

**Ministry of Natural Resources and Tourism**
Private Bag 7753
Causeway, Harare, Zimbabwe
794455

**Zimbabwe National Conservation Trust**
P.O. Box 8575
Causeway, Harare, Zimbabwe
70030

Non-government group.

# LAND

**Problem:** Only about 30% of the Earth's surface is covered with dry land, yet we continue to pollute and destroy this finite resource. Millions of trees are decimated daily, and our tainted soil produces contaminated food. Even lands, such as national parks and wilderness areas, are growing smaller.

**Solution:** Non-chemical alternatives to pesticides, organic farming, the halt of massive deforestation, and the worldwide replanting of the earth will help restore the land. Maintaining as much wild land as possible will remind us of what we have to lose.

This chapter includes government and non-government groups, international agencies, organizations, companies, and individuals mainly involved with: agriculture, caves, deserts, forestry, gardens, land use, mountains, national forests, outdoor recreation, plant conservation, rain forests, trails, and wilderness.

### Abundant Life Seed Foundation
P.O. Box 772
Port Townsend, WA 98368
(206) 385-5660
*Forest Shomer, Director*

Acquires, propagates, and protects the plants and seeds of the North Pacific Rim.

### Acres
1802 Chapman Road
Huntertown, IN 46748
(219) 637-6264
*Jane H. Dustin, Secretary*

Preserves the natural areas of northeastern Indiana.

### Adirondack Council
P.O. Box D-2
Elizabethtown, NY 12932
(518) 873-2240
*Barbara Glaser, Chairman*

Coalition of conservation organizations working to preserve the six-million-acre Adirondack Park.

### Adirondack Mountain Club
R.R. 3, Box 3055
Lake George, NY 12845
(518) 668-4447
*Joseph Dawson, President*

Promotes the "forever wild" concept in the Adirondack and Catskill mountains.

### Africa Tree Center
P.O. Box 90
Plessislaer, Natal, South Africa 20693

Founded in 1978 by black South Africans to combat soil erosion by tree planting.

### Alabama Department of Agriculture and Industries
Richard Beard Building, P.O. Box 3336
Montgomery, AL 36193
(205) 261-2550
*Albert McDonald, Commissioner*

## Alabama Forestry Commission
513 Madison Avenue
Montgomery, AL 36130
(205) 240-9300
*C. W. Moody, State Forester*

## Alberta Wilderness Association
Box 6398, Station D
Calgary, Alberta T2P 2E1 Canada
(403) 283-2025
*Vivian Pharis, President*

Lobbies for "non-motorized" outdoor experiences.

## Alliance for the Wild Rockies
415 N. Higgins Avenue
Missoula, MT 59802
(406) 721-5420

Coalition of environmentalists and business owners working to protect millions of acres of undeveloped land in the Rocky Mountains.

## Alternative Farming Systems Information Center
National Agriculture Library
Room 111
Beltsville, MD 20705
(301) 344-3724

Provides information and advice on how to farm and garden organically.

## American Association of Botanical Gardens and Arboreta
P.O. Box 206
Swarthmore, PA 19081
(215) 328-9145
*Hadley Osborne, President*
*William E. Barrick, Vice-President*

Established 1940.

## American Camping Association
Bradford Woods
5000 Street Road, 67N
Martinsville, IN 46151
(317) 342-8456
*Chuck Ackenbom, President*

Established 1910, preserves natural areas for camping and promotes careful use of campsites.

## American Cave Conservation Association
131 Main and Cave Streets,
P.O. Box 409
Horse Cave, KY 42749
(502) 786-1466
*David G. Foster, Executive Director*

Established in 1977 to preserve and protect caves.

## American Conservation Association
30 Rockefeller Plaza, Room 5402
New York, NY 10112
(212) 649-5600
*Laurance Rockefeller, Chairman of the Board*
*George R. Lamb, Executive Vice-President*

Advocates preservation and development of natural resources for public use.

## American Farm Bureau Federation
225 Touhy Avenue
Park Ridge, IL 60058
(312) 399-5700
*Dean Kleckner, President*

Deals with local, state, and national interests for its 3,500,000 member farmers. Established 1919.

Washington, DC
600 Maryland Avenue SW,
Suite 800
Washington, DC 20024
(202) 484-3600

## American Farmland Trust
1920 N Street, NW, Suite 400
Washington, DC 20036
(202) 659-5170
*Ralph Grossi, President*

Deals with soil erosion and the agricultural landbase.

**American Forest Council**
1250 Connecticut Avenue, NW,
Suite 320
Washington, DC 20036
(202) 463-2455
*Laurence D. Wiseman, President*

Supported by forest product manufacturers, develops productive management schemes for commercial forest lands. Established 1941.

Mid-Atlantic Region
415 River Street
Troy, NY 12181
(518) 272-0062
*John H. Herrington, Regional Manager*

New England Region
415 River Street
Troy, NY 12181
(518) 272-0062
*Jane Difley, Regional Manager*

Southern Region
2900 Chamblee Tucker Road,
Building 5
Atlanta, GA 30341
(404) 451-7106
*Mary Anne Lindskog, Regional Manager*

Western Region
1515 SW Fifth Avenue,
Suite 518
Portland, OR 97201
(503) 222-7456
*John E. Benneth, Regional Manager*

**American Forestry Association**
1516 P Street, NW
Washington, DC 20005
(202) 667-3300
*Richard M. Hollier, President*

Citizens' group whose Global ReLeaf program aims to plant trees to aid the environment. Established 1875.

**American Geographical Society**
156 Fifth Avenue, Suite 600
New York, NY 10010
(212) 242-0214
*John E. Gould, President*

Sponsors research projects and symposiums on man and the land.

**American Hiking Society**
1015 31st Street, NW
Washington, DC 20007
(703) 385-3252
*Charles Sloan, President*

Protects the interests of hikers and preserves American footpaths.

**American Recreation Coalition**
1331 Pennsylvania Avenue, NW,
#726
Washington, DC 20004
(202) 662-7420
*Derrick A. Crandall, President*
*David J. Humphreys, Chairman*

Federation of more than 100 recreation-related associations.

**American Society of
Agricultural Engineers**
2950 Niles Road
St. Joseph, MI 49085
(616) 429-0300
*Roger Castensen, Executive
Vice-President*

**American Society of Landscape
Architects**
4401 Connecticut Avenue, NW,
5th Floor
Washington, DC 20008
(202) 686-ASLA
*David Bohardt, Executive
Vice-President*

Authorized by the Department of Education to accredit landscape programs at colleges and universities.

# LAND

**American Trails**
1400 16th Street, NW
Washington, DC 20036
(202) 483-5611
*Charles Flink, Chairman*

Promotes support of hiking trails, planning, development, and maintenance. Founded in 1971.

**American Wild Lands Alliance**
7500 East Arapahoe Road,
Suite 355
Englewood, CO 80112
(303) 771-0380
*Sally A. Ranney, President*

National group for the protection and wise use of the remaining wilderness.

**Angeles National Forest**
701 Santa Anita Avenue
Arcadia, CA 91006
(818) 574-1613
*George A. Roby, Supervisor*

**Angelina, Davy Crockett, Sabine, and Sam Houston National Forests**
National Forests in Texas
Homer Garrison Federal Building
701 North 1st Street
Lufkin, TX 75901
(409) 639-8501
*William M. Lannan, Supervisor*

**Appalachian Mountain Club**
5 Joy Street
Boston, MA 02108
(617) 523-0636
*Earle Perkins, President*

Maintains trails and shelters, publishes guide books and maps. Founded 1876.

**Appalachian Regional Commission**
1666 Connecticut Avenue, NW
Washington, DC 20235
(202) 673-7893
*Jacqueline L. Phillips, Federal Co-Chairman*
*Gov. Robert P. Casey, States' Co-Chairman*

Provides a framework for federal and state efforts for all aspects of the Appalachian mountain range.

**Appalachian Trail Conference**
P.O. Box 807
Harpers Ferry, WV 25425
(304) 535-6331
*Margaret C. Drummond, Chairman*

Coordinates, preserves, and maintains the Appalachian Trail from Maine to Georgia. Founded 1925.

**Archeological Conservancy**
415 Orchard Drive
Santa Fe, NM 87501
(505) 982-3278
*Stewart L. Udall, Chairman*

Preserves prehistoric sites, also acquires significant land for archeological research.

**Arizona Commission of Agriculture and Horticulture**
1688 West Adams
Phoenix, AZ 85007
(602) 542-4373
*Ivan J. Shields, Director*

**Arizona/Sonora Desert Museum**
Route 9, Box 90
Tucson, AZ 85704
(602) 883-1380

Focuses on the natural resources of the Sonoran Desert.

## Arkansas Forestry Commission
P.O. Box 4523, Asher Station,
3821 West Roosevelt Road
Little Rock, AR 72214
(501) 664-2531
*James Bibler, Commission*
*Chairman*

## Big Thicket Association
Box 198, Highway 770
Saratoga, TX 77585
(409) 274-5000

Preserves the wilderness area of southeast Texas known as The Big Thicket, a 3,000,000-acre area that is a major habitat for migratory birds and endangered species.

## Big Thicket Conservation Association
P.O. Box 12032
Beaumont, TX 77706
(409) 892-8976
*Robert Finch, President*

Promotes appreciation and preservation of the natural and cultural heritage of southeastern Texas.

## Bioregional Project
HCR 3, Box 3
Brixey, MO 65618
(417) 679-4773
*David Haenke, Executive Officer*

A project of New Life Farm, working toward safe energy, organic agriculture, and forest husbandry.

## Black Hills National Forest
R.R. 1
Custer, SD 57730-9504
(605) 673-2251
*Darrel L. Kenops, Supervisor*

## Black Lung Association
907 West Neville Street
Beckley, WV 25801
(304) 253-2842
*Frank Hale, Treasurer*

Promotes mine health and safety standards.

## Boise Cascade
One Jefferson Square
Boise, ID 83728
(208) 384-6161
*Jon H. Miller, President*

Lumber/paper company in the forefront of reforestation and recycled paper.

## Bridger-Teton National Forest
Forest Service Building, 340 North Cache, Box 1888
Jackson, WY 83001
(307) 733-2752
*Brian Stout, Supervisor*

## California Department of Food and Agriculture
1220 North Street
Sacramento, CA 95814
(916) 445-9280
*Henry J. Voss, Director*

## California Department of Forestry and Fire Protection
1416 Ninth Street, P.O. Box 944246
Sacramento, CA 94244-2460
(916) 445-9920
*Richard J. Ernest, Director*

## Camp Fire Club of America
230 Camp Fire Road
Chappaqua, NY 10514
(914) 941-0199
*Thomas J. Fisher, President*

Committee on Conservation of Forests and Wildlife.

## Canadian Arctic Resources Committee
111 Sparks Street, Fourth Floor
Ottawa, Ontario K1P 5B5 Canada
(613) 236-7379

Concerned with the impact of Arctic development.

## Canadian Parks and Wilderness Society
69 Sherbourne Street, Suite 313
Toronto, Ontario M5A 3X7
Canada
(416) 366-3494

## Catskill Center for Conservation and Development
Arkville, NY 12406
(914) 586-2611
*William R. Ginsberg, President*

Works on land use planning and preservation of the Catskill Mountains.

## Cave Research Foundation
4074 West Redwing Street
Tucson, AZ 85741
(602) 744-2243
*Rondal R. Bridgeman, President*

Conducts research, interpretation, and conservation of caves. Helps conserve Carlsbad Caverns and Mammoth Caves.

## Center for Plant Conservation
125 Arborway
Jamaica Plain, MA 02130
(617) 524-6988
*Donald A. Falk, Executive Director*

Maintains a network of botanical gardens for rare and endangered plants.

## Chattahoochee and Oconee National Forests
508 Oak Street, NW
Gainesville, GA 30501
(404) 536-0541
*Kenneth D. Henderson, Supervisor*

## Chihuahuan Desert Research Institute
P.O. Box 1334
Alpine, TX 79831
(915) 837-8370
*James F. Scudday, President*

Researches, preserves, and educates concerning the Chihuahuan Desert of the southwest U.S. and northern Mexico.

## Children of the Green Earth
P.O. Box 95219
Seattle, WA 98145
(206) 781-0852
*Michael Soule, Board Member*
*Dorothy Craig, Board Member*

International organization committed to re-greening the earth through helping children plant trees.

## Chippewa National Forest
Cass Lake, MN 56633
(218) 335-2226
*William F. Spinner, Supervisor*

## Chugach National Forest
201 East Ninth Avenue, Suite 206
Anchorage, AK 99501
(907) 271-2500
*Dalton DuLac, Supervisor*

## Club of Friends of the Sahel
c/o CILSS, B.P. 7049
Ouagadougou, Burkina Faso

Concerned with ecological deterioration of area that borders the Sahara Desert.

## Coalition for Food Irradiation
10509 Walter Thompson Drive
Vienna, VA 22180
(703) 255-9195
*Nancy L. Blair, Executive Director*

Represents food trade associations promoting the use of food irradiation to preserve fruit and vegetables.

## Colorado Department of Agriculture
1525 Sherman Street
Denver, CO 80203
(303) 866-2811
*Naioma Benson, Chairman*

## Colorado Mountain Club
2530 West Alameda Avenue
Denver, CO 80219
(303) 922-8315
*Susan Baker, President*

Founded in 1912 to foster appreciation
of the mountains.

## Committee for National Arbor Day
640 Eagle Rock Avenue
West Orange, NJ 07052
(201) 731-0594
*Harry J. Banker, Chairman*

Organized in 1936 to establish the last
Friday in April as National Arbor Day.

## Connecticut Department of Agriculture
165 Capitol Avenue, Room 273
State Office Building
Hartford, CT 06106
(203) 566-4667
*Kenneth B. Andersen,
Commissioner*

## Continental Divide Trail Society
P.O. Box 30002
Bethesda, MD 20814
(301) 493-4080
*James R. Wolfe, Director*

Plans, develops, and maintains the
trail.

## Custer National Forest
P.O. Box 2556
Billings, MT 59103
(406) 657-6361
*Curtis W. Bates, Supervisor*

## Cycad Society
1161 Phyllis Court
Mountain View, CA 94040
(415) 964-7898
*David S. Mayo, Secretary-Treasurer*

Conserves and propagates the endangered cycad species (palm and fernlike
plants).

## Daniel Boone National Forest
100 Vaught Road
Winchester, KY 40391
(606) 745-3100
*Richard H. Wengert, Supervisor*

## Dasholi Gram Swarajya Mamdal (Chipko Movement)
P.O. Gopeshwar District, Chamoli
Uttar Pradesh 246401, India
83

Works to prevent the felling of trees
and promote environmental awareness
among the rural people of India.

## Dawes Arboretum
7770 Jacksontown Road, SE
Newark, OH 43055
(614) 323-2355
*Jo Dawes Higgins, Chairman*

Promotes the planting of forest and ornamental trees.

## Delaware Department of Agriculture
2320 South DuPont Highway
Dover, DE 19901
(302) 736-4811
*William B. Chandler, Jr., Secretary*

## Delaware Wild Lands
303 Main Street
Odessa, DE 19730
(302) 834-1332
*Holger H. Harvey, Director*

Acquires land on the Delmarva Peninsula for its natural resource value.

## Desert Protective Council
P.O. Box 4294
Palm Springs, CA 92263
(714) 779-2099
*Bill Neill, President*

Dedicated to the preservation of deserts.

## Elm Research Institute

Harrisville, NH 03450
(603) 827-3048
(800) FOR-ELMS
*John P. Hansel, Executive Director*

Creator of "Johnny Elmseed" to plant and propagate a disease-resistant American Liberty Elm.

## Emergency Earth Rescue Administration

1480 Hoyt Street, Suite 31
Lakewood, CO 80215
(303) 233-3548
*Philip Isely, President*

Environmental rescue organization, works to solve crises such as oil spills and chemical spills as quickly as possible with the least amount of environmental damage.

## Federation of Western Outdoor Clubs

365 West 29th
Eugene, OR 97401
(503) 686-1365
*Larry Cash, President*

Association of clubs for the enjoyment of and protection of scenic wilderness.

Northwest Conservation
Representative
1516 Melrose
Seattle, WA 98122

Washington, DC, Representative
645 Pennsylvania Avenue, SE
Washington, DC 20003

## Florida Department of Agriculture and Consumer Services

State Capitol
Tallahassee, FL 32399-0810
(904) 488-3022
*Doyle Conner, Commissioner*

## Florida Trail Association

P.O. Box 13708
Gainesville, FL 32604
(904) 378-8823
*Ethel Palmer, Acting Manager*

Maintains and develops the Florida Trail, a proposed 1,300-mile track between the Everglades and Pensacola.

## Forest Farmers Association

4 Executive Park East, P.O. Box 95385
Atlanta, GA 30347
(404) 325-2954
*Eley C. Frazier III, President*

Association of timberland owners in 15 southern states.

## Forest History Society

701 Vickers Avenue
Durham, NC 27701
(919) 682-9319
*William R. Sizemore, President*

Studies and records man's interaction with the forest environment.

## Forest Trust

P.O. Box 9238
Santa Fe, NM 87504-9238
(505) 983-8992
*Henry H. Carey, Director*

Seeks to improve and protect forest ecosystems.

## Forestry Canada/Forêts Canada

Place Vincent Massey, 351 St. Joseph Boulevard
Hull, Quebec K1A 1G5 Canada
(819) 997-1107
*Jean-Claude Mercier, Deputy Minister*

Maritimes Region
P.O. Box 400, Regent Street South
Fredericton, New Brunswick E3B 5P7 Canada
(506) 452-3500
*H. Oldham, Director-General*

Newfoundland and Labrador
Pleasantville, P.O. Box 6028
St. John's, Newfoundland A1C
5X8 Canada
(709) 772-6019
*J. Munro, Director-General*

Northwest Region
5320—122nd Street
Edmonton, Alberta T6H 3S5
Canada
(403) 435-7210
*A. D. Kill, Director-General*

Ontario Region
P.O. Box 490, 1210 Queen
Street, East
Saulte Ste. Marie, Ontario P6A
5M7 Canada
(705) 949-9461
*Dr. C. Winget, Director*

Pacific and Yukon Region
506 West Burnside Road
Victoria, British Columbia V8Z
1M5 Canada
(604) 388-0600
*J. Drew, Director-General*

Région du Québec
1055 rue du P.E.P.S., Case
Postale 3800
Ste-Foy, Quebec G1V 4C7
Canada
(408) 648-4991
*Yvan Hardy, Director-General*

**Friends of the Boundary Waters Wilderness**
1313 5th Street, SE, Suite 329
Minneapolis, MN 55414
(612) 379-3835
*Kevin Proescholdt, Executive Director*

Protects and preserves the Boundary Waters Canoe Area and the surrounding Quetico-Superior ecosystem.

**Friends of the Everglades**
202 Park Street
Miami, FL 33166
(305) 888-1230
*Marjory Stoneman Douglas, President*

Goal is "to foster and facilitate through education a harmonious co-existence between human and natural environment systems" in the Everglades.

**Friends of the Urban Forest**
512 2nd Street, 4th Floor
San Francisco, CA 94107
(415) 543-5000

Plants trees in the San Francisco Bay Area.

**Garden Club of America**
598 Madison Avenue
New York, NY 10022
(212) 753-8287
*Mrs. Edward A. Blackburn, Jr., President*

Publishes information packet, *The World Around You,* and is involved in wide range of conservation and beautification efforts, including billboard and sign control.

**Georgia Department of Agriculture**
19 Martin Luther King Drive
Capitol Square
Atlanta, GA 30334
(404) 656-3600
*Tommy Irvin, Commissioner*

**Georgia Forestry Commission**
Box 819
Macon, GA 31298-4599
(912) 744-3211
*John W. Mixon, Director*

Fosters and encourages re-forestation.

## Georgia-Pacific Corporation
133 Peachtree Street, NE
Atlanta, GA 30303
(404) 521-4000
*T. Marshall Hann, Jr., CEO*
Lumber/paper product company.

## Gila National Forest
2610 Silver Street
Silver City, NM 88061
(505) 388-8201
*David W. Dahl, Supervisor*

## Grassland Heritage Foundation
5460 Buena Vista
Shawnee Mission, KS 66205
(913) 677-3326
*Phillip S. Brown, President*

Acquires and preserves native prairie.

## Greater Yellowstone Coalition
P.O. Box 1874, 13 South Wilson
Bozeman, MT 59715
(406) 586-1593
*Thomas McNamee, President*

Seeks to preserve the ecosystem of Yellowstone National Park.

## Green Mountain Club
P.O. Box 889, 43 State Street
Montpelier, VT 05602
(802) 223-3463
*Brian T. Fitzgerald, President*

Built the Long Trail, a hiking trail from Massachusetts to the Canadian border.

## Green Mountain National Forest
Federal Building, 151 West Street,
 P.O. Box 519
Rutland, VT 05701-0519
(802) 773-0300
*Stephen C. Harper, Supervisor*

## Greensward Foundation
104 Prospect Park, W
Brooklyn, NY 11215
*Robert M. Makla, Director*

Works to improve the care of urban parks.

## Hardwood Research Council
P.O. Box 34518
Memphis, TN 38184-0518
(901) 377-1824
*John A. Pitcher, Director*

Research and education in forest management and utilization for lumber.

## Hawaii Department of Agriculture
P.O. Box 22159
Honolulu, HI 96822
(808) 548-7101
*Yukio Kitagawa, Chairperson*

## High Desert Museum
59800 South Highway 97
Bend, OR 97702
(503) 382-4754
*Robert W. Chandler, President of the Board of Trustees*

Living museum in portions of eight western states and British Columbia.

## Holly Society of America
304 Northwind Road
Baltimore, MD 21204
(301) 825-8133
*Lloyd C. Hahn, President*

Covers all aspects of "holly culture" (i.e., conservation, research, and hybridization).

## Idaho Department of Agriculture
P.O. Box 790
Boise, ID 83701
(208) 334-3240
*Richard R. Rush, Director*

**Illinois Department of Agriculture**
State Fairgrounds
Springfield, IL 62706
(217) 782-2172
*Larry A. Werries, Director*

**Illinois Nature Preserves Commission**
524 South Second Street, Lincoln Tower Plaza
Springfield, IL 62701-1787
(217) 785-8686
*G. Tanner Girard, Chairperson*

**Institute for Alternative Agriculture**
9200 Edmonston Road, Suite 117
Greenbelt, MD 20770
(301) 441-8777
*I. Garth Youngberg, Executive Director*

Advances systems of food and fiber production that are economically viable. Made up of organic and conventional farmers.

**International Alliance for Sustainable Agriculture**
Newman Center
University of Minnesota
1701 University Avenue, SE, Room 202
Minneapolis, MN 55414
(612) 331-1099
*Terry Gips, Founder*

This group works nationally and internationally to promote "economically viable, ecologically sound, socially just, and humane agricultural systems."

**International Centre for Integrated Mountain Development**
P.O. Box 3226, Jawalakhel, Lalitpur
Kathmandu, Nepal
521575

Works to promote economically and environmentally sound development of the Hindu Kush Himalayas.

**International Erosion Control Association**
P.O. Box 4904
Steamboat Springs, CO 80477
(303) 879-3010
*Ben Northcutt, Executive Director*

Provides opportunities for the exchange of information and ideas concerning effective and economical methods of erosion control.

**International Mountain Society**
P.O. Box 3128
Boulder, CO 80307
(303) 494-9228
*Jack D. Ives, Director*

Promotes international cooperation for protection of mountain land.

**International Paper**
Two Manhattanville Road
Purchase, NY 10577
(212) 536-6000
*John A. Georges, CEO*

Manufactures paper products.

**International Society for the Preservation of the Tropical Rainforest**
3931 Camino de la Cumbre
Sherman Oaks, CA 91423
(818) 788-2002

Promotes cooperation between the U.S. and the USSR.

**International Society of Arboriculture**
P.O. Box 908, 303 West University Avenue
Urbana, IL 61801
(217) 328-2032
*Michael Walterscheidt, President*

Preserves shade and ornamental trees.

## International Society of Tropical Foresters
5400 Grosvenor Lane
Bethesda, MD 20814
(301) 897-8720
*Warren T. Doolittle, President*

Objective is to transfer technology to natural resource managers and researchers in tropical forests.

## International Tropical Timber Organization
Sangyo Boeki Centre Building, 2 Yamashita-cho, Naka-ku
Yokohama 231, Japan
671 7045
671 7007 FAX

Industry group that works with tropical timber-producing and -consuming countries to expand trade.

## Iowa Department of Agriculture and Land Stewardship
Wallace State Office Building
Des Moines, IA 50319
(515) 281-5851
*James B. Gulliford, Director*

## Island Resources Foundation
Red Hook Center
Box 33
St. Thomas, Virgin Islands 00802
(809) 775-6225
*Edward L. Towle, President and Executive Director*

Protects the natural resources of islands.

Washington, DC
Webster House, Suite T-4, 1718 P Street, NW
Washington, DC 20036
(202) 265-9712

## Jackson Hole Preserve
30 Rockefeller Plaza, Room 5402
New York, NY 10112
(212) 649-5600
*Laurance S. Rockefeller, Chairman of the Board*

Conserves areas of outstanding natural beauty.

## Kansas State Board of Agriculture
109 S.W. 9th Street
Topeka, KS 66612
(913) 296-3556
*Sam Brownback, Secretary*

## Kentucky Department of Agriculture
Capitol Plaza Tower
Frankfort, KY 40601
(502) 564-4696
*Ward Burnette, Commissioner*

## Klamath National Forest
1312 Fairlane Road
Yreka, CA 96097
(916) 842-6131
*Robert L. Rice, Supervisor*

## Land Between the Lakes Association
100 Van Morgan Drive
Golden Pond, KY 42211-9001
(502) 924-5897
*Robert L. Herbst, Chairman*

Supports Tennessee Valley Authority's Land Between the Lakes, a 170,000-acre recreation and environmental education center.

## Land Trust Exchange
1017 Duke Street
Alexandria, VA 22314
(703) 683-7778
*Jean W. Hocker, President/Executive Director*

Builds awareness of the consequences of diminishing land resources.

**Landlab**
Cal-Poly Pomona
3801 West Temple Avenue
Pomona, CA 91768
(714) 869-3341
*Edwin A. Barnes III, Director*

Center for education and research in the sustainable use of natural resources.

**Landscape Journal**
The University of Wisconsin Press
114 North Murray Street
Madison, WI 53715
(608) 262-4928

Semi-annual publication.

**Law and the Land**
c/o Robinson and Cole
One Commercial Plaza
Hartford, CT 06103-3597
(203) 275-8200

Quarterly publication.

**Lewis and Clark National Forest**
Box 871, 1101 15th Street, N
Great Falls, MT 59403
(406) 791-7700
*J. Dale Gorman, Supervisor*

**Louisiana Department of Agriculture and Forestry**
P.O. Box 94302
Baton Rouge, LA 70804-9302
(504) 922-1234
*Bob Odom, Commissioner*

**Maine Department of Agriculture, Food, and Rural Resources**
State House Station #28
Augusta, ME 04333
(207) 289-3871
*Bernard W. Shaw, Commissioner*

**Mark Twain National Forest**
401 Fairgrounds Road
Rolla, MO 65401
(314) 364-4621
*B. Eric Morse, Supervisor*

**Maryland Department of Agriculture**
50 Harry S. Truman Parkway
Annapolis, MD 21401
(301) 841-5700
*Wayne A. Cawley, Jr., Secretary*

**Massachusetts Department of Food and Agriculture**
100 Cambridge Street
Boston, MA 02202
(617) 727-3002
*August Schumacher, Jr., Commissioner*

**Mead Corporation**
Courthouse Plaza, NE
Dayton, OH 45463
(513) 222-6323
*B. R. Roberts, CEO*

Manufactures forest products.

**Men of the Trees**
Turners Hill Road, Crawley Down
Crawley RH10 4HL England
712356

Works for large-scale tree planting worldwide.

**Men's Garden Clubs of America**
5560 Merle Hay Road
Johnston, IA 50131
(515) 278-0295
*Chris Christiansen, President*

Home gardeners organization with 150 chapters throughout the U.S.

**Michigan Department of Agriculture**
P.O. Box 30017
Lansing, MI 48909
(517) 373-1050
*Robert L. Mitchell, Director*

## Minnesota Department of Agriculture
90 West Plato Boulevard
St. Paul, MN 55107
(612) 297-2200
*James Nichols, Commissioner*

## Mississippi Department of Agriculture and Commerce
P.O. Box 1609
Jackson, MS 39215-1609
(601) 354-7050
*Jim Buck Ross, Commissioner*

## Missouri Department of Agriculture
P.O. Box 630, 1616 Missouri Boulevard
Jefferson City, MO 65101
(314) 751-4211
*Charles E. Kruse, Director*

## Missouri Prairie Foundation
P.O. Box 200
Columbus, MO 65205
(816) 361-1700
*Jerry Overton, President*

Protects the native prairie.

## Monongahela National Forest
USDA Building, Sycamore Street, Box 1548
Elkins, WV 26241-1548
(304) 636-1800
*Jim Page, Supervisor*

## Montana Department of Agriculture
Environmental Management Division
Agriculture and Livestock Building, Capitol Station
Helena, MT 59620-0201
(406) 444-3144
*Everett Snortland, Director*

## Montana Land Reliance
P.O. Box 355
Helena, MT 59624
(406) 443-7027
*Allen Bjergo, President*

Protects and preserves ecologically and agriculturally significant land.

## Montana Outfitters and Guides Association
P.O. Box 631
Hot Springs, MT 59845
(406) 741-2811
*Rhoda G. Cook, Executive Director*

Outfitters and guides who operate wilderness trips in Montana for hunting, fishing, and sightseeing. Encourages preservation of back country and wise use of resources.

## Montana Wilderness Association
P.O. Box 635
Helena, MT 59624
(406) 443-7350
*Lou Bruno, President*

Preserves and manages public lands.

## Mountaineers, The
300 Third Avenue, West
Seattle, WA 98119
(206) 284-6310
*Carsten Lien, President*

Explores, studies, and preserves the Northwest.

## National Arbor Day Foundation
100 Arbor Avenue
Nebraska City, NE 68410
(402) 474-5655
*Susan Seacrest, President*

Sponsors Trees for America, Arbor Day, Tree City USA, etc.

## National Association of State Departments of Agriculture

1616 H Street, NW
Washington, DC 20006
(202) 628-1566
*Arthur R. Brown, President*

Includes the heads of departments of agriculture.

## National Association of State Foresters

c/o Forest, Park and Wildlife Services
580 Taylor Avenue
Annapolis, MD 21501-2351
(301) 269-3776
*James B. Roberts, President*

## National Association of State Park Directors

c/o Parks and Recreation Department
1424 West Century Avenue, Suite 202
Bismarck, ND 58501
(701) 224-4887
*Douglas K. Eiken, President*

## National Association of State Recreation Planners

205 Butler Street, SE, Suite 1352
Atlanta, GA 30334
(404) 656-2753
*Terri Yearwood, President*

## National Coalition to Stop Food Irradiation

P.O. Box 59-0488
San Francisco, CA 94159
(408) 626-2734

Information clearinghouse and speakers' bureau against the use of irradiation of food.

## National Council of State Garden Clubs

4401 Magnolia Avenue
St. Louis, MO 63110
(314) 776-7574
*Mrs. C. Manning Smith, President*

Preserves civic beauty.

## National Council of the Paper Industry for Air and Stream Improvement

260 Madison Avenue
New York, NY 10016
(212) 532-9000
*Isaiah Gellman, President*

Conducts research on environmental problems related to industrial forestry and the manufacture of pulp, paper, and wood products.

## National Farmers Union

Denver, CO 80251
(303) 337-5500
*Leland H. Swenson, President*

Goal is to preserve farmlands for future generations.

## National Future Farmers of America Organization

P.O. Box 15160. National FFA Center
Alexandria, VA 22309
(703) 360-3600
*Larry D. Case, Advisor*

National organization of high school agriculture students.

## National Gardening Association

180 Flynn Avenue
Burlington, VT 05401
(802) 863-1308
*Charles Scott, President*

Helps people become successful gardeners at home.

## National Grange, The
1616 H Street, NW
Washington, DC 20006
(202) 628-3507
*Robert E. Barrow, Master*

Oldest farmer's organization in the U.S. Founded in 1867.

## National Inholders Association
30 West Thomson, Box 588
Sonoma, CA 95476
(707) 996-5334
*Charles S. Cushman, Executive Director*

Association of persons holding property on the boundaries of federal lands or holding property impacted by federal lands or preserves.

## National Parks and Conservation Association
1015 31st Street, NW
Washington, DC 20007
(202) 944-8530
*Norman G. Cohen, Chairman of Executive Committee*

Preserves, promotes and improves the nation's parks. (For a complete listing of National Parks, see the U.S. Department of the Interior, National Park Service, pages 83–86.)

## National Recreation and Park Association
3101 Park Center Drive
Alexandria, VA 22302
(703) 820-4940
*Kathryn A. Porter, Chairman of the Board*

Improves park and recreation leadership.

## National Save the Family Farm Coalition
80 F Street, NW
Washington, DC 20001
(202) 737-2215
*Benny Bunting, President*

Advocates sustainable agriculture.

## National Speleological Society
Cave Avenue
Huntsville, AL 35810
(205) 852-1300
*John Scheltens, President*

Explores, studies and preserves caves.

## National Tropical Botanical Gardens
P.O. Box 340
Lawai, Kauai, HI 96765
(808) 332-7361
*William Theobald, Director*

Protects tropical plants from all over the world.

## National Wilderness Institute
25766 Georgetown Station
Washington, DC 20007
(703) 836-7404
*Robert E. Gordon, Jr., Director*

Protects all wilderness areas in the U.S.

## National Wildflower Research Center
2600 FM 973 North
Austin, TX 78725
(512) 929-3600
*David Northington, Executive Director*

Founded by Mrs. Lyndon B. Johnson, promotes the preservation of native plants.

## National Woodland Owners Association
374 Maple Avenue, E, Suite 204
Vienna, VA 22180
(703) 255-2700
*Bert W. Udell, Chairman*

**Native Plant Society of Texas**
P.O. Box 891
Georgetown, TX 78627
(512) 863-7794
*Sarah Wasowski, President*

Concerned with conservation and preservation of Texas' native plants.

**Native Seeds/Search**
3950 West New York Drive
Tucson, AZ 85745
(602) 327-9123
*Gary Nabhan, President*

Conserves traditional crop seeds and related wild seeds of the American Southwest.

**Natural Areas Association**
320 South Third Street
Rockford, IL 61104
(815) 964-6666
*Glenn Juday, President*

Provides a channel for the exchange of information on natural areas preservation.

**Nebraska Department of Agriculture**
301 Centennial Mall South
Lincoln, NE 68509
(402) 471-2341
*George Beattie, Director*

**Nevada Department of Agriculture**
350 Capitol Hill Avenue
Reno, NV 89510
(702) 789-0180
*Thomas W. Ballow, Executive Director*

**New England Forestry Foundation**
85 Newbury Street
Boston, MA 02116
(617) 437-1441
*Hugh Putnam, Jr., Executive Director*

Owns and manages more than 15,000 acres of forest in five states, including educational centers on sound forestry practices for private woodland owners.

**New England Trail Conference**
33 Knollwood Drive
East Longmeadow, MA 01028
(413) 525-7052
*Forrest E. House, Secretary-Treasurer*

Includes groups in six New England states and parts of New York who are interested in hiking, trail clearing, and maintenance.

**New England Wild Flower Society**
Garden in the Woods
Grenway Road
Framingham, MA 01701
(617) 237-4924
*Geraldine Payne, President*

**New Forests Project**
731 Eighth Street, SE
Washington, DC 20003
*Lindsay Mattison, Executive Director*

Plants multi-purpose trees in the Third World to combat hunger, poverty, and environmental deterioration.

**New Hampshire Department of Agriculture**
Caller Box 2042
Concord, NH 03302
(603) 271-3551
*Stephen H. Taylor, Commissioner*

**New Jersey Agricultural Society**
CN 331
Trenton, NJ 08625
(609) 394-7766
*Edward V. Lipman, President*

Founded in 1781 to improve and promote agriculture in New Jersey.

## New Jersey Department of Agriculture

CN 330
Trenton, NJ 08625
(609) 292-5530
*Arthur R. Brown, Secretary*

## New Jersey Pinelands Commission

P.O. Box 7
New Lisbon, NJ 08064
(609) 894-9342
*Richard Sullivan, Chairman*

State planning and regulatory agency with jurisdiction over land use and development in the Pinelands region.

## New Mexico Department of Agriculture

P.O. Box 30005, Department 3189
Las Cruces, NM 88003-0005
(505) 646-3007
*Frank A. Dubois, Director and Secretary of Agriculture*

## New York Adirondack Park Agency

P.O. Box 99
Ray Brook, NY 12977
(518) 891-4050
*Herman F. Cole, Jr., Chairman*

Develops state and private land use for the 6,000,000-acre Adirondack Park.

## New York Department of Agriculture and Markets

1 Winners Circle, Capitol Plaza
Albany, NY 12235
(518) 457-3880
*Richard T. McGuire, Commissioner*

## New York-New Jersey Trail Conference

232 Madison Avenue
New York, NY 10016
(212) 685-9699
*H. Neil Zimmerman, President*

Builds and maintains 800 miles of foot trails.

## North American Family Campers Association

P.O. Box 328
Concord, VT 05824
(802) 695-2563
*Harold Coakely, President*

Encourages camping without harming the environment.

## North American Farm Alliance

P.O. Box 2502
Ames, IA 50010
(515) 232-1008
*Merle Hanson, President*

An organization of farmers concerned with land use and food prices.

## North Carolina Department of Agriculture

P.O. Box 27647
Raleigh, NC 27611
(919) 733-7125
*James A. Graham, Commissioner*

## North Dakota Department of Agriculture

600 East Boulevard Avenue, 6th Floor
Bismarck, ND 58505-0020
(701) 224-2231
*Sarah Vogel, Commissioner*

## Northeastern Forest Fire Protection Commission

10 Ladybug Lane
Concord, NH 03301
(603) 224-6966
*Richard E. Mullavey, Executive Director*

Coordinates mutual training and cooperation between seven U.S. states and two Canadian provinces.

## Northern Plains Resource Council

419 Stapleton Building
Billings, MT 59101
(406) 246-1154
*Teresa Erickson, Staff Director*

Association of resource agencies in the Plains states.

**Northwest Renewable Resources Center**
1133 Dexter Horton Bldg.,
710 Second Avenue
Seattle, WA 98104
(206) 623-7361
*James C. Waldo, Chairman*

Formed by industry leaders, Indian tribes, and environmental organizations to offer alternative dispute resolution services on natural resource issues.

**Ohio Department of Agriculture**
65 South Front Street
Columbus, OH 43215
(614) 466-2732
*Steven D. Maurer, Director*

**Oklahoma State Board of Agriculture**
2800 North Lincoln Boulevard
Oklahoma City, OK 73105
(405) 521-3864
*Jack Craig, President*

**Olympic National Forest**
Box 2288
Olympia, WA 98507
(206) 753-9534
*Ted C. Stubblefield, Supervisor*

**Olympic Park Associates**
13245 40th Avenue, NE
Seattle, WA 98125
(206) 364-3933
*Polly Dyer, President*

Works to preserve Olympic National Forest.

**Open Space Institute**
40 West 20th Street
New York, NY 10011
(212) 949-1966
*Thomas Whyatt, Executive Director*

Protects open space and promotes environmental education.

**Oregon Department of Agriculture**
Salem, OR 97310-0110
(503) 378-3773
*Dalton Hobbs, Director,*
*Information Service*

**Oregon State Department of Forestry**
2600 State Street
Salem, OR 97310
(503) 378-2560
*James E. Brown, State Forester*

**Outdoors Unlimited**
P.O. Box 373
Kaysville, UT 84037
(801) 544-0960
*Marlene Simons, President*

Individuals who believe that public lands should be used for the benefit of the American people and that natural resources can be used without harming the environment.

**Ozark and St. Francis National Forests**
605 West Main, Box 1008
Russellville, AR 72801
(501) 321-5202
*Lynn C. Neff, Supervisor*

**Ozark Society**
P.O. Box 2914
Little Rock, AR 72203
*Stewart Noland, President*

Promotes knowledge and enjoyment of the Ozark-Ouachita mountain range.

**Partners in Parks**
1855 Quarley Place
Henderson, NV 89014
(702) 454-5547
*Sarah G. Bishop, Chair and*
*President*

Individuals and organizations who support the national parks.

## Pennsylvania Department of Agriculture
2301 North Cameron Street
Harrisburg, PA 17110-9408
(717) 787-4737
*Boyd E. Wolff, Secretary*

## People, Food and Land Foundation
35751 Oaks Springs Drive
Tollhouse, CA 93667
(209) 855-3710
*George Ballis, Coordinator*

Small farmers concerned with low water use, arid land crops, solar models, and organic farming.

## Piedmont Environmental Council
P.O. Box 460
Warrenton, VA 22186
(703) 347-2334
*Charles S. Whitehouse, Chairman*

Conserves the natural resources and landscape of the Piedmont region in northern Virginia.

## Potomac Appalachian Trail Club
1718 N Street, NW
Washington, DC 20036
(202) 638-5306
*Warren Sharp, President*

Maintains 240 miles of the Appalachian Trail.

## Prairie Club, The
Suite 603A, 10 S. Wabash Avenue
Chicago, IL 60603
(312) 236-3342
*John Lonk, President*

Watchdog group for the remaining American prairie.

## Preserving Family Lands
P.O. Box 2242
Boston, MA 02107
(617) 244-7553

Magazine dealing with private land ownership issues.

## Public Lands Foundation
P.O. Box 10403
McLean, VA 22102
(703) 790-1988
*George D. Lea, President*

Preserves public lands.

## Rails-to-Trails Conservancy
1400 Sixteenth Street, NW, Suite 300
Washington, DC 20036
(202) 797-5400
*David G. Burwell, President*

Wants the public to use trains rather than cars to travel into national parkland.

## Rainforest Action Network
301 Broadway, Suite A
San Francisco, CA 94133
(415) 398-4404
*Randall, Hayes, Director*

Supports strong measures to stop the destruction of the world's rainforests.

## Rainforest Alliance
270 Lafayette Street, Suite 512
New York, NY 10012
(212) 941-1900
*Daniel R. Katz, President*

Seeks to expand awareness of the role the U.S. plays in the fate of the tropical rainforests.

## Revkin, Andrew
c/o Houghton Mifflin
One Beacon Street
Boston, MA 02108
(617) 725-5000

Author of *The Burning Season: The Murder of Chico Mendes and the Fight for the Amazon Rain Forest.*

## Santa Monica Mountains Conservancy
107 South Broadway, Room 7117
Los Angeles, CA 90017
(213) 620-2021
*Joseph Edmiston, Director*

Responsible for the protection of the Santa Monica Mountains.

## Save the Dunes Council
444 Barker Road
Michigan City, IN 46360
(219) 879-3937
*Thomas Serynek, President*

Preserves and protects lake-shore dunes.

## Save the Rainforest Action Committee
P.O. Box 34427
Los Angeles, CA 90034
(213) 281-1907

## Save-the-Redwoods League
114 Sansome Street, Room 605
San Francisco, CA 94104
(415) 362-2352
*John B. DeWitt, Executive Director*

Preserves representative stands of coast and sierra redwoods.

## Scenic America
216 7th Street, SE
Washington, DC 20003
(202) 546-1100
*Carroll Shaddock, President*

Fights billboard blight.

## Scott Polar Research Institute
Lensfield Road
Cambridge CB2 1ER, England
36541

A leading center of research on polar zones.

## Sequoia National Forest
900 West Grand Avenue
Porterville, CA 93257
(209) 784-1500
*James A. Crates, Supervisor*

## Sierra National Forest
1130 O Street, Room 3009
Fresno, CA 93721
(209) 487-5155
*James L. Boynton, Supervisor*

## Siskiyou National Forest
Box 440
Grants Pass, OR 97526
(503) 479-5301
*Ronald J. McCormick, Supervisor*

## Society for Ecological Restoration and Management
University of Wisconsin Arboretum
1207 Seminole Highway
Madison, WI 53711
(608) 263-7889

Works to develop public support for maintaining natural areas in urban, rural, and wilderness settings.

## Society for Range Management
1839 York Street
Denver, CO 80206
(303) 355-7070
*Peter V. Jackson, Executive Vice-President*

Studies rangeland resources for wildlife and livestock.

## Society for the Protection of New Hampshire Forests
54 Portsmouth Street
Concord, NH 03301-5400
(603) 224-9945
*Malcolm McLane, Chairman*

Organized in 1901 to promote the wise use of renewable resources.

## Society of American Foresters
5400 Grosvenor Lane
Bethesda, MD 20014
(301) 897-8720
*William H. Banzaf, Executive Vice-President*

Professional society of foresters and scientists working in related fields.

## South Carolina Department of Agriculture
Wade Hampton Office Building
P.O. Box 11280
Columbia, SC 29211
(803) 734-2210
*D. Leslie Tindal, Commissioner*

## South Carolina Forestry Commission
Box 21707
Columbia, SC 29221
(803) 737-8800
*Robert J. Gould, State Forester*

## South Carolina Land Resources Conservation Commission
2221 Devine Street, Suite 222
Columbia, SC 29205
(803) 734-9100
*William S. Simpsom III, Chairman*

## South Dakota Department of Agriculture
445 East Capitol, Sigurd Anderson Building
Pierre, SD 57501
(605) 773-3375
*Jay C. Swisher, Secretary*

## Stanislaus National Forest
19777 Greenley Road
Sonora, CA 95370
(209) 532-3671
*Blaine L. Cornell, Supervisor*

## Tennessee Citizens for Wilderness Planning
130 Tabor Road
Oak Ridge, TN 37830
(615) 482-2153
*Martha Ketelle, President*

Develops plans concerning wilderness and natural areas and acts with government and private citizens to establish policies.

## Tennessee Department of Agriculture
P.O. Box 40627, Melrose Station, Ellington Agricultural Ctr.
Nashville, TN 37204
(615) 360-0103
*A. C. Clark, Commissioner*

## Tennessee Forestry Association
P.O. Box 290693
Nashville, TN 37229
(615) 883-3832
*Don Kettenbeil, President*

Trade group made up of woodland owners, public and private foresters, educators, and wood-using companies.

## Texas Agricultural Extension Service
Texas A&M University
College Station, TX 77843
(409) 845-7967
*Zerle L. Carpenter, Director*

Provides information regarding farming and soil conservation to farmers.

## Texas Department of Agriculture
P.O. Box 12847, Capitol Station
Austin, TX 78711
(512) 463-7476
*Jim Hightower, Commissioner*

**Texas Forestry Association**
P.O. Box 1488
Lufkin, TX 75901
(409) 632-TREE
*Bob Currie, President*

Promotes conservation and economic development of forests. Established 1914.

**Texas General Land Office**
Stephen F. Austin State Office
    Building
Austin, TX 78701
(512) 463-5256
*Garry Mauro, Commissioner*

Custodian of state-owned land.

**Thomas Cook Group**
45 Berkeley Street
London W1A 1EBB, England
071-4994000
*John Brooks, Chairman*

Ecologically sensitive tours.

**Tongass-Chatham Area
    National Forest**
204 Siginaka Way
Sitka, AK 99835
(907) 747-6671

**Tongass-Ketchikan Area
    National Forest**
Federal Building
Ketchikan, AK 99901
(907) 225-3101
*Mike Lunn, Supervisor*

**TreePeople**
12601 Mulholland Drive
Beverly Hills, CA 90210
(213) 753-4600
*Andy Lipkis, Founder*
*Katie Lipkis, Vice-President*

Raises funds for and plants trees in parkland and in cities.

**Trees For Life**
1103 Jefferson
Wichita, KS 67203
(316) 263-7294

Has planted a million trees in India and is organizing a program to help plant 100 million food-bearing trees in underdeveloped countries.

**Trees for Tomorrow**
611 Sheridan Street, P.O. Box 609
Eagle River, WI 54521
(715) 479-6456
*Henry H. Haskell, Executive
    Director*

Organization of paper mills, power companies and other industries seeking to conserve forests.

**Trust for Public Land**
116 New Montgomery Street, 4th
    Floor
San Francisco, CA 94105
(415) 495-4014
*Martin J. Rosen, President*

Dedicated to acquiring and preserving land in urban and rural areas for parks and gardens.

**Trustees of Reservations**
572 Essex Street
Beverly, MA 01915
(508) 921-1944
*Herbert W. Vaughan, Chairman*

Massachusetts land conservation organization founded in 1891.

## Utah Department of Agriculture
350 North Redwood Road
Salt Lake City, UT 84116
(801) 538-7100
*Miles (Cap) Ferry, Commissioner*

## Utah Nature Study Society
323 South 2nd West
Tooele, UT 84074
(801) 882-4729
*Eldon Romney, President*
*Maria Dickerson, Secretary*

Promotes conservation and nature education through workshops and field trips.

## Utah Wilderness Association
455 East 400 South, #306
Salt Lake City, UT 84111
(801) 359-1337
*Dick Carter, Coordinator*

Organization working on all land issues in Utah.

## Vermont Department of Agriculture
116 State Street
Montpelier, VT 05602
(802) 828-2430
*Ronald A. Allbee, Commissioner*

## Virginia Department of Agriculture
Consumer Services
P.O. Box 1163
Richmond, VA 23209
(804) 786-3501
*S. Mason Carbaugh, Commissioner*

## Virginia Forestry Association
1205 E. Main Street
Richmond, VA 23219
(804) 644-8462
*Robert Lundberg, President*

Association of groups, individuals, industries, clubs, and businesses formed to promote the wise use and development of Virginia's forests.

## Virginia Native Plant Society
P.O. Box 844
Annandale, VA 22003
(703) 368-9803
*Nicky Staunton, President*

Promotes the conservation of wild plants and habitats.

## Walden Forever Wild
P.O. Box 275
Concord, MA 01742
(203) 429-2839
*Mary P. Sherwood, Chairman*

Objective is to change Walden Pond Reservation from an recreational park to an educational, historical, and ecological sanctuary.

## Walden Pond Advisory Committee
Page Road
Lincoln, MA 01773
(617) 259-9544
*Kenneth E. Bassett, Chairman*

Representatives of communities and organizations interested in preserving and restoring the Walden Pond area.

## Wasatch-Cache National Forest
8230 Federal Building, 125 South State Street
Salt Lake City, UT 84138
(801) 524-5030
*Dale Bosworth, Supervisor*

**Washington Department of Agriculture**
406 General Administration
Building
Olympia, WA 98504
(206) 753-5063
*C. Alan Pettibone, Director*

**Washington Farm Forestry Association**
P.O. Box 7663
Olympia, WA 98507
(206) 459-0984
*Norman Hutson, Jr., President*

Affiliated with National Woodland Owners Association.

**Washington State Forestry Conference**
College of Forest Resources
University of Washington
Seattle, WA 98195
(206) 545-4960
*David B. Thorud, President*

Provides public forum for public and private foresters.

**West Virginia Department of Agriculture**
State Capitol Building
Charleston, WV 25305
(304) 348-3550
*Cleve Benedict, Commissioner*
*Ralph Glover, State Forester*

**West Virginia Geological and Economic Survey**
Box 879
Morgantown, WV 26507-0879
(304) 594-2331
*Larry D. Woodfork, Director and State Geologist, Administration*

**Western Forestry and Conservation Association**
4033 SW Canyon Road
Portland, OR 97221
(503) 226-4562
(503) 228-3624 FAX
*Jim Bentley, President*

Forest landowners, forestry associations, and individual foresters from eight western states and Canada.

**Weyerhaeuser Company**
Tacoma, WA 98477
(206) 924-2345
*George A. Weyerhaeuser, CEO*

Forest products manufacturer.

**Wilderness Society**
1400 I Street, NW, Suite 550
Washington, DC 20005
(202) 842-3400
*Alice Rivlin, Chair*
*George Frampton, Jr., President*

Purpose is to establish a "land ethic" through education of the public. Founded in 1935.

Alaska Region
519 West 8th Avenue, Suite 205
Anchorage, AK 99501
(907) 272-9453
*Allen Smith, Director*

California and Nevada Region
116 New Montgomery, Suite 526
San Francisco, CA 94105
(415) 541-9144
*Patricia Schifferle, Director*

Central Rockies Region (CO, UT)
777 Grant Street, Suite 606
Denver, CO 80206
(303) 839-1175
*Darrel Knuffke, Director*

Florida Keys
8065 Overseas Highway, Suite 4
Marathon, FL 33050
(305) 289-1010
*Ross Burnaman, Director*

Florida Region (FL, PR, VI)
4203 Ponce de Leon Boulevard
Coral Gables, FL 33146
(305) 448-3636
*James D. Webb, Director*

Inter-Mountain (MT, WY, ID)
413 West Idaho Street, Suite 102
Boise, ID 83702
(208) 343-8153
*Craig Gehrke, Director*

Northeast Region (MA, VT,
NH, ME, CT, RI, NJ, NY)
20 Park Plaza, #536
Boston, MA 02116
(617) 350-8866
*Michael Kellett, Director*

Northern Rockies (MT, WY)
105 West Main Street, Suite E
Bozeman, MT 59715
(406) 586-1600

Northwest Region (WA, OR)
1424 Fourth Street, Room 816
Seattle, WA 98101
(206) 624-6430
*Jean Duirning, Director*

Southeast Region (AL, AR, GA,
KY, LA, MS, NC, OK, SC,
TN, East TX, WV)
1819 Peachtree Road, NE,
Suite 714
Atlanta, GA 30309
(404) 355-1783
*Peter Kirby, Director*

Southwest Region (AZ, NM,
West TX)
234 North Central Avenue
Phoenix, AZ 85004
(602) 256-7921
*James Norton, Director*

## Wilderness Watch
P.O. Box 782
Sturgeon Bay, WI 54235
(414) 743-1238
*Jerome O. Gandt,
President-Treasurer*

Dedicated to the sustained use of America's sylvan lands.

## Willamette National Forest
Box 10607
Eugene, OR 97440
(503) 687-6521
*Michael A. Kerrick, Supervisor*

## Wisconsin Department of Agriculture
801 West Badger Road
Madison, WI 53713
(608) 267-9788

## Woman's National Farm and Garden Association
2402 Clearview Drive
Glenshaw, PA 15116
(412) 486-7964
*Rita H. Kirschler, President*

Stimulates interest in the conservation of natural resources.

## World Forestry Center
4033 SW Canyon Road
Portland, OR 97221
(503) 228-1367

Studies trends in world forestry.

**World Rainforest Movement**
87 Cantonment Road
10250 Penang, Malaysia
4 373713
*Martin Khor Kok Peng, Director*

**Wyoming Department of
Agriculture**
2219 Carey Avenue
Cheyenne, WY 82002
(307) 777-7321
*Don Rolston, Commissioner*

**Wyoming Forestry Division**
1100 West 22nd Street
Cheyenne, WY 82002
(307) 777-7586
*Bryce E. Lundell, State Forester*

# MEDIA AND CELEBRITIES

This chapter includes book publishers, film and video producers, individuals (from journalists to authors to movie and television stars), magazines, media organizations, newsletters, newspapers, and television networks either involved in the environmental movement or acting as important outlets for public information about environmental issues.

We encourage everyone to write to media outlets about their coverage of environmental issues and make them aware of local environmental concerns.

**ABC Television Network**
77 West 66th Street
New York, NY 10023
(212) 456-1000
*John B. Sias, President*

**Allegheny Press**
19323 Elgin Road
Corry, PA 16407
(814) 664-8504
*Dr. John Tomikel, Publisher*

Specializes in publishing books and pamphlets on environmental problems.

**Ban the Box**
c/o The KITA Group
12 East 41st Street, Suite 1600
New York, NY 10017
(212) 684-2550

Group of record companies and musicians striving to eliminate the long box packaging for compact discs in order to save forests and prevent waste.

**Bardot, Brigitte**
La Madrique, Aix-en-Provence
Saint Tropez, France

Actress/environmental activist, particularly in causes for animals.

**Bazell, Robert**
c/o NBC News
30 Rockefeller Plaza
New York, NY 10112
(212) 664-4444

Science television journalist.

**Begley, Ed, Jr.**
8899 Beverly Boulevard
Los Angeles, CA 90048

Actor/environmental activist who uses public conveyances or bicycle for transportation.

**Berlet Films**
1646 Kimmel Road
Jackson, MI 49201
(517) 784-6969

Sells and rents environmental films and publishes a free catalog.

**Bernhard, Sandra**
10100 Santa Monica Boulevard,
  Suite 1600
Los Angeles, CA 90069

Comedienne who works with Conservation International.

**Better World Society**
1100 17th Street NW, Suite 502
Washington, DC 20036
(202) 331-3770
*Ted Turner, Founder*
*Tom Belford, Executive Director*

Produces and acquires solution-oriented television programming on the environment.

**Bullfrog Films**
Oley, PA 19547
(800) 543-3764

Rents and sells films to educational institutions.

**Bureau of National Affairs, Inc.**
1231 25th Street, NW
Washington, DC 20037
(202) 452-4200
*William A. Beltz, President*

Publishes *Environment Reporter.*

**Buzzworm**
1818 Sixteenth Street
Boulder, CO 80302

Photography-oriented environmental magazine.

**Cable News Network (CNN)**
100 International Boulevard
Atlanta, GA 30348
(404) 827-1700
*Ted Turner, Chairman*

**Canadian Nature Federation**
46 Elgin Street
Ottawa, Ontario K1P 5K6 Canada

Publishes *Nature Canada* magazine.

**CBS Television Network**
51 West 52nd Street
New York, NY 10019
(212) 975-4321
*Lawrence Alan Tisch, Chairman*

**Chamberlain, Richard**
9830 Wilshire Boulevard
Beverly Hills, CA 90212

Actor who works with American Rivers.

**Cher**
9200 Sunset Boulevard, #1001
Los Angeles, CA 90069

Singer/actress and co-founder of Mothers and Others for the Environment.

**Consumer Information Center**
P.O. Box 100
Pueblo, CO 81002

Publishes consumer information for the U.S. government.

**Contemporary Films/McGraw Hill Eastern Region**
Princeton/Hightstown Road
Hightstown, NJ 08520

Environmental films/film strips available.

**Corporation for Public Broadcasting**
1111 16th Street, NW
Washington, DC 20036
(202) 955-5100
*Howard D. Grutin, President*

The financing arm for the Public Broadcasting Service (PBS).

**Denver, John**
P.O. Box 1587
Aspen, CO 81611

Singer who's helped found the Aspen Institute on Global Change.

**Earth Communications Office**
1925 Century Park East, Suite 2300
Los Angeles, CA 90067
(213) 662-5207
*Bonnie Reiss, Executive Director*

An environmental group that educates and marshals the resources of celebrities and the entertainment industry.

## Earthscan
3 Endsleigh Street
London WC1H ODD, England
01 388 2117
*Neil Middleton, Managing Director*

Publishing house for environmental subjects.

## Earthwise Consumer
P.O. Box 1506
Mill Valley, CA 94942
(415) 383-5892
*Debra Lynn Dodd, Editor*

Newsletter on everyday products and earth-wise topics.

## Earthworks Press
1400 Shattuck Avenue
Berkeley, CA 94709
(415) 841-5866
*John Javna, Owner*

Publishers of environmentally oriented consumer books.

## Ecology Law Quarterly
University of California
Boalt Hall School of Law
Berkeley, CA 94720
(415) 642-0457
*Richard Allan, Editor*

## Educational Communications
P.O. Box 35473
Los Angeles, CA 90035
(213) 559-9160
*Nancy Pearlman, Editor and Executive Producer*

Publishes the *Compendium Newsletter,* a guide to ecological activism.

## Environment Communications
6410 Rockledge Drive, Suite 203
Bethesda, MD 20817
(301) 571-9791

Publishes *Congressional Directory: Environment.*

## Environmental Film Review
Environment Information Center
Film Reference Department
292 Madison Avenue
New York, NY 10017
Films and catalog available.

## Environmental Film Service
National Association of
Conservation Districts
408 East Main, P.O. Box 855
League City, TX 77573
(713) 332-3402

Makes available to the public films on the environment and conservation.

## Environmental Images, Inc.
300 I Street, NW, Suites 100 & 325
Washington, DC 20036
(202) 675-9100

Films, videotapes, and slides available.

## Environmental Media Association
10536 Culver Boulevard
Culver City, CA 90232
(213) 559-9334
*Andy Spahn, President*

Information clearinghouse created to make entertainment industry producers and writers aware of environmental issues and information.

## Environmental Outlook
Institute for Environmental Studies
University of Washington, FM012
Seattle, WA 98185

Monthly publication.

## Environmental Science and Technology
American Chemical Society
1155 16th Street, NW
Washington, DC 20036
(202) 872-4600

Monthly publication.

**Field and Stream Magazine**
380 Madison Avenue
New York, NY 10017
(212) 779-5000
*Duncan Barnes, Editor*

Broad-based magazine for the hunter and fisherman.

**Films and Research for an Endangered Environment**
201 North Wells, Suite 1735
Chicago, IL 60606
(312) 782-1376

Film presentation available upon request.

**Fonda, Jane**
P.O. Box 491355
Los Angeles, CA 90049

Actress and environmental activist.

**Forum International: International Ecosystems University**
91 Gregory Lane, #21
Pleasant Hill, CA 94523
(415) 671-2900
*Nicolas D. Hetzer, Director*

Publishes bi-monthly *Ecosphere,* which describes "ecosystemic whole-world-oriented, transdisciplinary, value-based education, research and action programs."

**Gabriel, Peter**
25 Ives Street
London SW3 England

First major recording star to insist on eliminating the paper-wasteful longbox for CD packaging; also a Greenpeace supporter.

**Georgia Department of Natural Resources**
Film Unit
270 Washington Street
Atlanta, GA 30334

Will loan environmental films as well as sell them. Free catalog available upon request.

**Hawn, Goldie**
c/o Hollywood Pictures
500 S. Buena Vista Avenue
Burbank, CA 91505

Actress and co-founder of Mothers and Others for the Environment.

**Heldref Publications**
4000 Albermarle Street, NW, Suite 504
Washington, DC 20016
(202) 362-6445

Publishes *Environment* and *The Journal of Environmental Education.*

**Heloise**
P.O. Box 795000
San Antonio, TX 78279
(512) HELOISE FAX

Domestic engineering tipster and author of *Hints For A Healthy Planet.*

**Henley, Don**
10880 Wilshire Boulevard, Suite 2110
Los Angeles, CA 90024

Musician who works with Defenders of Wildlife.

**Herman, Pee Wee**
P.O. Box 48243
Los Angeles, CA 90048

Comic actor who works with Greenpeace USA.

**Heston, Charlton**
8730 Sunset Boulevard, 6th Floor
Los Angeles, CA 90069

Actor/world population activist.

**High Country News**
Box 1090
Paonia, CO 81428
(303) 527-4898
*Betsy Marston, Editor*

Bi-weekly publication specializing in mountain environment issues.

**Hollender, Jeffrey**
c/o William Morrow and
  Company
105 Madison Avenue
New York, NY 10016
(212) 889-3050

Author of *How to Make the World a Better Place.*

**Hynes, H. Patricia**
c/o Prima Publishing &
  Communications
P.O. Box 1260PH
Rocklin, CA 95677
(916) 624-5718

Author of *Earth Right.*

**In Business**
JG Press
Box 351
Emmaus, PA 18049
(215) 967-4135
*Nora Goldstein, Executive Editor*

Bi-weekly business publication, publishes articles on environmentally concerned businesses.

**International Film Bureau**
332 South Michigan Avenue
Chicago, IL 60604
(312) 427-4545

Free catalog for sales and rental of films on the environment.

**Island Press**
1718 Connecticut Avenue, NW,
  Suite 300
Washington, DC 20009
(202) 232-7933

Publishing company specializing in environmental publications.

**Izaak Walton League of
  America**
Publishing Division
1800 North Kent Street, Suite 806
Arlington, VA 22209

Publishes *Outdoor America.*

**JG Press**
Box 323
Emmaus, PA 18049
(215) 967-4135

Publishing house specializing in environmental issues.

**Lamb, Marjorie**
c/o Harper & Row
10 East 53rd Street
New York, NY 10022
(212) 207-7000

Author of *2 Minutes a Day for a Greener Planet.*

## Levine, Michael
8730 Sunset Boulevard
Los Angeles, CA 90069

Author of *The Environmental Address Book.*

## Los Angeles Times
Times Mirror Company
Times Mirror Square
Los Angeles, CA 90053
(213) 237-5000
*David Laventhol, Publisher*

## Madonna
9200 Sunset Boulevard, #915
Los Angeles, CA 90069

Singer/actress who works with Conservation International.

## Media Network
121 Fulton Street, 5th Floor
New York, NY 10038
(212) 619-3455

Computerized database of thousands of films, videotapes, and slide shows, including environmental subjects.

## Midler, Bette
P.O. Box 46703
Los Angeles, CA 90046

Actress/singer who works with Mothers and Others for the Environment.

## Midnight Oil
P.O. Box 186, Glebe
Sydney 2037 New South Wales,
    Australia

Environmentally concerned rock group. Leader Peter Garrett is president of the Australian Conservation Foundation.

## Modern Talking Picture Service
500 Park Street, North
St. Petersburg, FL 33709
(813) 541-6763

Catalog includes environmental films.

## Mother Earth News
P.O. Box 70
Hendersonville, NC 28793
(704) 693-0211
*Bruce Woods, Editor*

Bi-monthly publication emphasizing country living and country skills.

## Mother Jones Magazine
1663 Mission Street
San Francisco, CA 94103
(415) 558-8881
*Richard Reynolds, Communications Director*

Publishes articles on environmental concerns.

## MTV Networks
1775 Broadway
New York, NY 10019
(212) 484-8680
*Tom Freston, Chairman*

Active in encouraging environmental awareness through its programming and support of Greenpeace.

## Muppets, The
(Kermit, Miss Piggy, the Cookie Monster, Big Bird, etc.)
117 E. 69th Street
New York, NY 10021
(212) 794-2400

Spokespuppets for the National Wildlife Federation.

## Natural Resources Journal
University of New Mexico School
of Law
1117 Stanford, NE
Albuquerque, NM 87131
(505) 277-2146

Quarterly publication.

## NBC Television Network
30 Rockefeller Plaza
New York, NY 10020
(212) 664-4444
*Robert C. Wright, CEO*

## New York Times
229 West 43rd Street
New York, NY 10036
(212) 556-1234
*Arthur Ochs Sulzberger, Publisher*

## Newsweek
444 Madison Avenue
New York, NY 10022
(212) 350-4000
(212) 350-4120 FAX (Letters to
the Editor)
*Richard M. Smith, Editor-in-Chief*

## Newton-John, Olivia
P.O. Box 2710
Malibu, CA 90265

Singer and honorary ambassador,
United Nations Environment Programme.

## Null, Gary
c/o Villard Books
201 East 50th Street
New York, NY 10022
(212) 751-2600

Author of *Clearer, Cleaner, Safer,
Greener.*

## Olin, Ken
222 North Canon Drive, Suite 222
Beverly Hills, CA 90210

Actor associated with the Sierra Club.

## Omni Magazine
1965 Broadway
New York, NY 10023
(212) 496-6100
*Patrice Adcroft, Editor*

## One Person's Impact
P.O. Box 751
Westborough, MA 01581
(508) 478-3716
*Sarah Blair/Maria Valenti,
Spokespersons*

Newsletter with tips to help save the
environment.

## Outdoor Canada Magazine
801 York Mills Road, Suite 301
Don Mills, Ontario M3B 1X7,
Canada
(416) 443-8888
(416) 443-1869 FAX
*Teddi Brown, Editor-in-chief*

Canadian outdoor recreation magazine.

## Outdoor Writers Association of America
2017 Cato Avenue, Suite 101
State College, PA 16801
(814) 234-1011
*Sylvia G. Bashline, Executive
Director*

Professional organization of staff and
freelance writers concerned with outdoor recreation and conservation.

**Outside**
1165 North Clark Street
Chicago, IL 60610
(312) 951-0990
*Lawrence J. Burke, Publisher and Editor-in-chief*

Magazine geared to outdoor pursuits, heavy concentration on environmental concerns.

**Pollution Abstracts**
Cambridge Scientific Abstracts
7200 Wisconsin Avenue
Bethesda, MD 20814
(301) 961-6700

**Pollution Control Guide**
Commerce Clearing House, Inc.
4025 West Peterson Avenue
Chicago, IL 60646
(312) 583-8500

**Rachel Carson Homestead Association**
613 Marion Avenue
Springdale, PA 15144
(412) 443-8792
*Joseph R. Panza, President*

Persons and organizations interested in carrying on the principles of Rachel Carson and who maintain the house and grounds of her birthplace.

**Redford, Robert**
P.O. Box 837
Provo, UT 84601

Actor and environmental activist, particularly in the American West.

**Rodale Press and Research Center**
33 East Minor Street
Emmaus, PA 18098
(215) 967-5171

Publishes *The New Farm, Organic Gardening,* and *Prevention* magazines. Also published the book *The Chemical Free Lawn.*

**Rolling Stone Magazine**
745 5th Avenue
New York, NY 10151
(212) 758-3800
*Jann S. Wenner, Editor*

Youth magazine with a high concentration of environmentally oriented articles.

**Sheen, Martin**
P.O. Box 4293
Malibu, CA 90265

Actor/environmental activist, anti-nuclear spokesman.

**Sierra Club**
Periodical Office
530 Bush Street
San Francisco, CA 94108
(415) 981-8634

Publishes weekly *National News Report* and bi-monthly *Sierra* magazine.

**Sierra Club Books**
2034 Fillmore Street
San Francisco, CA 94115
(415) 776-2211
*Daniel Moses, Editor-in-chief*

Book publishing division of the Sierra Club.

**Starke, Linda**
c/o Oxford University Press
200 Madison Avenue
New York, NY 10016
(212) 679-7300

Author of *Signs of Hope,* good news on the condition of the environment.

**Stewart, Jimmy**
8899 Beverly Boulevard
Los Angeles, CA 90048

Actor associated with African Wildlife Foundation.

**Streep, Meryl**
P.O. Box 105
Taconic, CT 06079

Actress and co-founder of Mothers and Others for the Environment.

**Television Trust for the Environment**
46 Charlotte Street
London W1P 1LX England
637 4602
580 7780 FAX

Distribution service for environmental films in Great Britain.

**Tiegs, Cheryl**
7060 Hollywood Boulevard, Suite 1010
Hollywood, California 90028

World-famous model who works with the Wilderness Society.

**Time Magazine**
Time & Life Building, Rockefeller Center
New York, NY 10020-1393
(212) 522-1212
(212) 522-0601 FAX (Letters to the Editor)
*Jason McManus, Editor-in-Chief*

**Trends Publishing**
National Press Building
Washington, DC 20045
(202) 393-0031

Publishes *Environment Report* semi-monthly.

**USA Today**
Gannett Publishing
1000 Wilson Boulevard
Arlington, VA 22229
(703) 276-3400
*Thomas J. Farrell, General Manager*

**Vallely, Bernadette**
c/o Ivy Books/Ballantine Books
201 E. 50th Street
New York, NY 10022
(212) 751-2600

Author of *1,001 Ways To Save The Planet.*

**Washington Post**
1150 15th Street, NW
Washington, DC 20071
(202) 334-6000
*Donald Graham, Publisher*

**West Wind Productions**
P.O. Box 3532
Boulder, CO 80307
(303) 443-2800

Distributes 16mm films and videos on environmental topics.

**Whole Earth Catalog**
Portola Institute
Menlo Park, CA 94025
(415) 845-3000
*Stewart Brand, Founder*

Catalog of environmental products, organizations and issues.

**Williams, Robin**
9830 Wilshire Boulevard
Beverly Hills, CA 90212

Actor who works with Mothers and Others for the Environment.

# POPULATION

**Problem:** A rapidly increasing world population depletes natural resources such as air, land and water; accelerates waste problems; and causes economic hardships which make solutions to environmental problems ever more difficult.

**Solution:** Population must be sustainable—by the land it occupies, the food it produces, and the resources it develops. Many groups advocate birth control as a means to controlling population growth.

This chapter includes groups who are involved with: birth control, family planning, quality of life, and other aspects of population stabilization.

**Association for Voluntary Surgical Contraception**
122 East 42nd Street
New York, NY 10168
(212) 351-2500
*Joseph E. Davis, M.D., Chairman*
*Janet H. Halpern, Manager of*
*Public Information*

Established 1943, advocates sterilization as means of birth control and population stabilization.

**Baulieu, Etienne-Emile**
Hospital de Bicetre
94 Bicetre, France

Developer of RU-486, the "abortion pill."

**Carrying Capacity**
1325 G Street, NW, Suite 1003
Washington, DC 20005
(202) 879-3045
*Linda Kovan, Executive*
*Administrator*

Concerned with the "carrying capacity" of the earth, which is defined as the number of organisms that resources can supply indefinitely without degradation.

**Center for Communication Programs**
527 St. Paul Place
Baltimore, MD 21202
(301) 659-6300
*Phyllis T. Piotrow, Director*

Promotes and develops public awareness of family planning, fertility control, and world population balance.

**Centre for Development and Population Activities**
1717 Massachusetts Avenue, NW, #202
Washington, DC 20036
(202) 667-1142
*Kaval Gulhati, President*

Supports population professionals from Third World countries by providing training and technical assistance.

## Environmental Balance
1325 G Street, NW, Suite 1003
Washington, DC 20005
(202) 879-3000
*M. Rupert Cutler, Executive Director*

Promotes stabilization of population.

## Global Committee of Parliamentarians on Population and Development
304 East 45th Street, 12th Floor
New York, NY 10017
(212) 953-7947
*Akio Matsumura, Executive Director*

Seeks balance of population and resources to stabilize the world environment.

## Inter-American Parliamentary Group on Population and Development
902 Broadway, 10th Floor
New York, NY 10010
(212) 995-8860
*Rebeca Rios, Executive Secretary*

Legislators from North and South America who promote an improvement in the quality of life in the Western Hemisphere.

## International Planned Parenthood Federation
Regent's College, Regent's Park
London NW1 4NS England
486 0741
487 7950 FAX

## Morrison Institute for Population and Resource Studies
Stanford University
Stanford, CA 94305
(415) 723-2300

## Negative Population Growth
16 East 42nd Street, Suite 1042
New York, NY 10017
(212) 599-2020
*Donald Mann, President*

Promotes a drastic decrease in the world population to half of the present birth levels as "the only viable option consistent with human survival."

## Planned Parenthood Federation of America
810 Seventh Avenue
New York, NY 10019
(212) 541-7800
*Faye Wattleton, President*

Promotes availability of effective means of birth control in the U.S. and worldwide.

## Population Communication
1489 East Colorado Boulevard
Pasadena, CA 91106
(818) 793-4750
*Robert Gillespie, President*

Promotes global population stabilization.

## Population Concern
231 Tottenham Court Road
London W1P OHX England
631 1546

Provides public education and fundraising for population programs worldwide.

## Population Council
1 Dag Hammarskjold Plaza
New York, NY 10017
(212) 644-1300
*George Zeidenstein, President*

Works to support human reproductive biomedical sciences aimed at the development and improvement of birth control methods.

## Population Crisis Committee

Main Office, Suite 550, 1120 19th
Street, NW
Washington, DC 20036
(202) 659-1833
*J. Joseph Sepidel, President*

Encourages those activities that promise the greatest impact on reducing world population growth.

## Population Institute

110 Maryland Avenue, NE
Washington, DC 20002
(202) 544-3300
*Werner Fornos, President*

Seeks to marshall public opinion on global over-population problems.

## Population Reference Bureau

777 14th Street, NW, Suite 800
Washington, DC 20005
(202) 639-8040
*Thomas Merrick, President*

Gathers and disseminates information on world population trends.

## Population Renewal Office

36 West 59th Street
Kansas City, MO 64113
(816) 363-6980
*Frances Frech, Director*

Advocates population growth and opposes population control.

## Population Resource Center

622 Third Avenue
New York, NY 10017
(212) 888-2820
*Jane S. DeLung, President*

Provides demographic analyses of population for policy makers.

## Population-Environment Balance, Inc.

1325 G Street, NW, Suite 1003
Washington, DC 20005
(202) 879-3000
*David F. Durham, Chairman of the Board*

Promotes understanding of the relationship between population balance/distribution and the nation's well-being.

## Program for the Introduction and Adaptation of Contraceptive Technology

Four Nickerson Street
Seattle, WA 98109
(206) 285-3500
*Gordon W. Perkin, President*

Works to promote the availability of family planning in developing countries.

## United Nations Population Fund

220 East 42nd Street
New York, NY 10017
(212) 850-5631

Largest internationally funded source of population assistance worldwide. More than a quarter of the assistance to developing countries goes through this agency.

## World Federation of Health Agencies for the Advancement of Voluntary Surgical Contraception

122 East 42nd Street
New York, NY 10168
(212) 351-2526
*Beth S. Atkins, Executive Director*

Works to include voluntary surgical contraception as a choice within basic health services.

## World Population Society

1333 H Street, Suite 760
Washington, DC 20005
(202) 898-1303
*Frank Oram, Executive Director*

Initiates research about the impact of population change on our capacity to meet human needs.

## Zero Population Growth

1400 16th Street, NW, Suite 320
Washington, DC 20036
(202) 332-2200
*Susan Weber, Executive Director*

Veteran population control organization founded in 1968.

# RECYCLING AND HAZARDOUS WASTE

Problem: Landfills and dumps are being filled to capacity, and we are running out of places to dump trash. Chemicals are contaminating our food, water, and air. Solution: Reuse, recycle, reduce. Produce less waste and recycle the waste we do have. Eliminate non-biodegradable substances which clog landfills. Strictly regulate hazardous waste disposal.

This chapter includes government and non-government groups, international agencies, organizations, companies, and individuals mainly involved with: hazardous, nuclear, solid and general waste disposal; recycling of aluminum, garden waste, glass, paper, plastic, tin and steel; and re-usable products.

**Aluminum Association, The**
900 19th Street, NW, Suite 300
Washington, DC 20006
(202) 862-5100
*John C. Bard, President*

Manufacturers' organization.

**Aluminum Company of America (ALCOA)**
1501 Alcoa Building
Pittsburgh, PA 15219
(412) 553-4545
*Paul H. O'Neill, CEO*

Producer that promotes recycling of aluminum cans.

**Aluminum Recycling Association**
1000 16th Street, NW
Washington, DC 20036
(202) 785-0951
*Richard M. Cooperman, Executive Director*

Producers of aluminum alloys from scrap.

**American Paper Institute**
260 Madison Avenue
New York, NY 10016
(212) 340-0600
*Red Cavaney, President*

Manufacturers of pulp paper.

**Baby Bunz & Company**
P.O. Box 1717
Sebastopol, CA 95473
(707) 829-5347

Makes re-usable, pre-shaped cloth diapers.

**Biobottoms Fresh Air Wear**
P.O. Box 6009
Petaluma, CA 94953
(707) 778-7945

Makes pre-folded velour re-usable diapers.

**Browning-Ferris Industries**
14701 St. Mary's Lane
Houston, TX 77079
(713) 870-8100
*Harry J. Phillips, Chairman of the Board*

Primary business is the disposal of solid and chemical waste.

## California Waste Management Board
1020 Ninth Street, Suite 300
Sacramento, CA 95814
(916) 322-3330
*John E. Gallagher, Chairman*

## Center for Hazardous Materials Research
University of Pittsburgh Applied Research Center
320 William Pitt Way
Pittsburgh, PA 15238
(412) 826-5320
*Edgar Berkey, President*

Seeks to develop practical solutions to the problems associated with hazardous waste management.

## Center for Plastics Recycling Research
Rutgers University
Building 3529, Busch Campus
Piscataway, NJ 08855

Publishes manuals on collecting and sorting disposed plastic products. Also operates a plastics recycling plant.

## Central States Resource Center
809 South Fifth
Champaign, IL 61820
(217) 344-2371
*John W. Thompson, Executive Director*

Concerned with water policy and the recycling of solid, hazardous, and radioactive waste.

## Clean Sites, Inc.
1199 North Fairfax Street
Alexandria, VA 22314
(703) 683-8522
*Thomas Grumbly, President*

Mediates between parties at cleanup sites and publishes *Making Superfund Work.*

## Clean World International
c/o Keep Britain Tidy Group
Bostel House, 37 West Street
Brighton BN1 2RE, England
23585

Coalition of anti-littering groups and recycling organizations.

## Coalition for Responsible Waste Incineration
1330 Connecticut Avenue, NW, Suite 300
Washington, DC 20036
(202) 659-0060
*William S. Murray, Director*

Promotes responsible handling and incineration of industrial waste.

## Community Environmental Council
930 Miramonte Drive
Santa Barbara, CA 93109
(805) 963-0583
*Paul Relis, Director*

Advises local governments and businesses on recycling. Publishes *Gildea Review.*

## Conservatree Paper Company
10 Lombard Street, #250
San Francisco, CA 94111
(800) 522-9200
*Susan Kinsella, Spokesperson*

Sells recycled paper products and helps develop new markets for such products.

## Council for Solid Waste Solutions
1275 K Street, NW
Washington, DC 20006
(202) 371-5319
*Richard P. Swigart, Director of Communications*

A plastics industry group interested in the waste and recycling of their products.

**Dano Enterprises**
75 Commercial Street
Plainview, NY 11803
(516) 349-7300

Produces Ecolobags, an alternative to plastic garbage bags.

**Do Dream Music**
P.O. Box 5623
Takoma Park, MD 20912
(301) 445-3845

Has created *Recyclemania,* a cassette tape of songs by Billy B. to teach children to recycle.

**Earth Care Paper Company**
P.O. Box 3335
Madison, WI 53704
(608) 256-5522
*John Schaeffer, Founder*

Manufactures recycled paper products.

**Ecology Center**
2530 San Pablo Avenue
Berkeley, CA 94702
(415) 548-2220
*Karen Pickett, President*

Specializes in public education on environmental alternatives, particularly recycling.

**Environmental Action Coalition**
625 Broadway
New York, NY 10012
(212) 677-1601
*Nancy A. Wolf, Executive Director*

Focuses on source-separation recycling and monitors resource recovery installations.

**Environmental Management Association**
1019 Highland Avenue
Largo, FL 34640
(813) 586-5710
*Harold C. Rowe, President*

Administrators of environmental sanitation maintenance programs in industrial plants.

**Free-Flow Packaging**
1093 Charter Street
Redwood City, CA 94063
(415) 364-1145
*Arthur Graham, Founder*

Recycles polystyrene foam packaging if you send it to him (no CODs).

**Garbage: The Practical Journal for the Environment**
P.O. Box 56520
Boulder, CO 80321-6520
(303) 447-9330
*Patricia Poore, Editor*

Bi-monthly magazine.

**Geosafe**
Battelle Pacific Northwest
  Laboratories
4000 NE 41st
Seattle, WA 98105
(206) 525-3130
*James Hansen, Spokesperson*

Has developed a hazardous waste disposal process using glassification.

**Glass Packaging Institute**
1801 K Street, NW, Suite 1105L
Washington, DC 20006
(202) 887-4850
*Lewis D. Andrews, Jr., President*

Promotes the manufacture, use, and recycling of glass containers and closures.

**Governmental Refuse Collection and Disposal Association**
8750 Georgia Avenue, Suite 123,
  P.O. Box 7219
Silver Spring, MD 20910
(301) 585-2898
*H. Lanier Hickman, Jr., Executive Director*

Public agency officials and private corporate officials whose goal is to improve solid waste management services.

## H. T. Berry Company
50 North Street
Canton, MA 02021
(617) 828-6000

Distributes a full line of recycled products, including toilet tissue, paper towels, and napkins.

## Hackensack Meadowlands Development Commission
Environment Center Museum
2 DeKorte Park Plaza
Lyndhurst, NJ 07071
(201) 460-8300

Has the Tunnel of Trash, a simulated landfill, to help educate people on the reality of waste issues.

## Hazardous Materials Control Research Institute
9300 Columbia Boulevard
Silver Spring, MD 20910
(301) 587-9390
*Harold Bernard, Executive Director*

Disseminates technical information and promotes the use of risk assessment methods to achieve a balance between industrial growth and the environment.

## Hazardous Waste Federation
c/o New Mexico Hazardous Waste
  Management Society
Division 3314, P.O. Box 5800
Albuquerque, NM 87185
(505) 846-2655
*Gordon J. Smith, Chairman*

Seeks to increase public awareness and understanding of the problems related to hazardous waste management and to ensure protection of the environment.

## Hazardous Waste Treatment Council
1440 New York Avenue, NW,
  Suite 310
Washington, DC 20005
(202) 783-0870
*Richard C. Fortuna, Executive Director*

Interested in using high technology to dispose of waste in the interest of public health and safety.

## Institute for Local Self-Reliance
2425 18th Street, NW
Washington, DC 200009
(202) 232-4108
*Neil Seldman, Director*

Published *The United States Recycling Movement,* 1968 to 1986. Major projects include urban waste management.

## Institute of Scrap Recycling Industries
1627 K Street, NW
Washington, DC 20006
(202) 466-4050
*Herschel Cutler, Executive Director*

Processors, brokers, and consumers engaged in the recycling of metallic and non-metallic scrap.

## Iowa Waste Reduction Center
University of Northern Iowa
Cedar Falls, IA 50614
(319) 273-2079
*John Konefes, Director*

Matches companies producing waste in Iowa with others that use waste as a raw material.

## Kaiser Aluminum and Chemical
300 Lakeside Drive
Oakland, CA 94612
(415) 271-3300
*Cornel C. Maier, Chairman*

## Keep America Beautiful
9 West Broad Street
Stamford, CT 06902
(203) 323-8987
*Roger W. Powers, President*

Encourages a behaviorally based approach to recycling and waste issues. Founded 1953.

## National Association of Diaper Services

2017 Walnut Street
Philadelphia, PA 19103
(215) 569-3650
*John A. Shiffert, Executive Director*

Encourages the use of cloth, re-usable diapers rather than disposable ones.

## National Association of Solvent Recyclers

1333 New Hampshire Avenue, NW, Suite 1100
Washington, DC 20036
(202) 463-6956
*Barbara B. Wells, Director*

Seeks responsible recycling and reclamation of used industrial solvents.

## National Environmental Health Association

720 South Colorado Boulevard, South Tower, #970
Denver, CO 80222
(303) 756-9090
*Nelson E. Fabian, Executive Director*

Professional society of sanitarians.

## National Food and Conservation Through Swine

Fox Run Road, R.R. 4, Box 297
Sewell, NJ 08080
(609) 468-5447
*Ronnie Polen, Secretary*

Food waste collectors and feeders of swine who claim to be the oldest recycling industry in the U.S., since more than 6,000,000 tons of food are recycled through swine.

## National Recycling Coalition

17 M Street, NW, Suite 294
Washington, DC 20036
(202) 659-6883
*David Loveland, Executive Director*

Encourages recycling in businesses.

## National Resource Recovery Association

1620 I Street, NW
Washington, DC 20006
(202) 293-7330
*Ronald W. Musselwhite, Executive Secretary*

Encourages development of recycling programs and urban waste-to-energy systems.

## National Solid Wastes Management Association

1730 Rhode Island Avenue, NW, Suite 1000
Washington, DC 20036
(202) 659-4613
*Eugene J. Wingerter, Executive Director*

Publishes *Recycling: Treasure in Our Trash.* Maintains speakers' bureau, compiles statistics, and conducts research programs.

## Native Americans for a Clean Environment

307 S. Muskogee Avenue
Tahlequah, OK 74464
(918) 652-6298

Seeks to eliminate toxic and radioactive waste on Native American land.

## Natural Baby Company

R.D. 1, Box 160 S
Titusville, NJ 08560
(609) 737-2895

Manufactures re-usable cloth diapers.

## Ocean Arks International

1 Locust Street
Falmouth, MA 02540
(508) 540-6801

Developing solar aquatic waste treatment plants.

## Paper Service, Ltd.

P.O. Box 45
Hinsdale, NH 03451
(603) 239-4934

Manufactures recycled napkins and toilet paper.

## Pennsylvania Resources Council
P.O. Box 88
Media, PA 19063
(215) 565-9131
*Robert G. Struble, Jr., President*

Experts in recycling, publishes the *Environmental Shopping Guide.*

## Pollution Control Industries
One Fairfield Crescent
West Caldwell, NJ 07006
(201) 575-7052
*Larrie S. Calverts, President*

Produces ozone generators and specializes in waste disposal.

## Recoup
P.O. Box 577
Ogdensburg, NY 13669
(800) 267-0707

Lists 14,000 companies that purchase recyclable materials in *American Recycling Market* magazine.

## Recycled Paper Company
185 Corey Road
Boston, MA 02146
(617) 277-9901

Offers a full line of copier paper, stationery, envelopes, computer paper, and printing paper.

## Recycled Paper Products, Inc.
3636 North Broadway
Chicago, IL 60613
(312) 348-6410
*Mike Keiser, Co-founder*

## Recycling Legislation Action Coalition
177 Winthrop Road, Apartment 1
Brookline, MA 02416
(617) 232-9038
*Richard D. Wimberly, Executive Director*

Promotes and enforces recycling of solid materials.

## Resource Recycling
P.O. Box 10540
Portland, OR 97210
*Steve Apotheker, Technical Editor*

Publishes *Resource Recycling: North America's Recycling Journal.*

## Save A Tree
P.O. Box 862
Berkeley, CA 94701
(415) 526-9032

Sells washable canvas bags for shopping.

## Seattle Tilth Association
4649 Sunnyside Avenue, North
Seattle, WA 98103

Experts in home composting, publishes how-to brochure.

## Society of the Plastics Industry
1275 K Street, NW, Suite 400
Washington, DC 20005
(202) 371-5200
*Larry C. Thomas, President*

Publishes a list of companies that recycle plastic waste.

## Steel Can Recycling Institute
680 Anderson Drive
Pittsburgh, PA 15220
(800) 876-7274
*Kurt Smalberg, President*

Provides information and technical analyses to steel companies on methods of collection, preparation, and transportation of steel metal scrap.

## Thermal Energy Management Company
15 Woodside Avenue
Metuchen, NJ 08840
(201) 494-9198
*Richard H. Schlossel, Spokesperson*

Hazardous waste cleanup specialists.

**TOXNET**
Specialized Information Services
Division
National Library of Medicine
8600 Rockville Pike
Bethesda, MD 20894
(301) 496-6531

A database system that leads callers step-by-step through the process of cleaning up toxic spills.

**Waste Management, Inc.**
3003 Butterfield
Oak Brook, IL 60521
(312) 242-4317

Major hazardous waste and refuse hauling company.

**Waste Watch**
P.O. Box 39185
Washington, DC 20016
(202) 475-1684

Published the book *Waste Watcher.*

**Wellman, Inc.**
1040 Broad Street, #302
Shrewsbury, NJ 07702
(201) 542-7300

Largest plastics recycler in the U.S.

**Windstar Foundation**
2317 Snowmass Creek Road
Snowmass, CO 81654
(800) 669-4777

Publishers of "Recycling: 101 Practical Tips for Home and Work" booklet.

# TECHNOLOGY AND MONEY

**Problem:** At present, there is no all-encompassing effort to merge innovative scientific talents, research funding, government initiative, and the business community to develop and utilize the new technologies that can solve our environmental problems.

**Solution:** Financial resources ought to be marshaled from private and public sources to fund projects which have broad support from government, business, and citizens. A united effort from these concerned sources will speed up the development of new technology and help to solve our environmental problems.

This chapter includes government and non-government groups, international agencies, organizations, companies, and individuals mainly involved with: developing new technology; economic issues; financing environmentally conscious corporations, companies, and projects; job opportunities in environmental fields; providing technical and consulting assistance; and the sale and distribution of new, environmentally safe products.

**Agency for International Development (AID)**
Washington, DC 20523
(202) 647-1850
*Ronald W. Roskens, Director*

Principal U.S. government agency for international development assistance.

**America the Beautiful Fund**
219 Shoreham Building
Washington, DC 20005
(202) 638-1649
*Joshua Peterfreund, President*
*Paul Bruce Dowling, Secretary*

Provides recognition, technical support, and grants of seed money for local action groups beautifying America. Established 1965.

**American Academy of Environmental Medicine**
P.O. Box 16106
Denver, CO 80216
(303) 622-9755
*J. A. Howard, Executive Administrator*

Physicians, engineers, and others interested in the clinical aspects of environmental medicine.

**American Association for the Advancement of Science**
1333 H Street, NW
Washington, DC 20005
(202) 326-6400
*Donald N. Langengerg, President*
*Richard C. Atkinson, Chairman of the Board*

Fosters cooperation, scientific freedom, and responsibility among scientists.

## American Council on the Environment

1301 20th Street, NW, Suite 113
Washington, DC 20036
(202) 659-1900
*John H. Gullett, Executive Officer*

Organization of business and professional people, unions, and others interested in the environment as well as the economic well-being of the nation.

## American Institute of Architects

1735 New York Avenue, NW
Washington, DC 20006
(202) 626-7300
*James P. Cramer, CEO*

Promotes environmentally sound architectural practices.

## American Institute of Biological Sciences

730 11th Street, NW
Washington, DC 20001-4584
(202) 628-1500
*Paul R. Ehrlich, President*

National organization for biology and biologists, established 1947. Publishes *Bio Science Forum.*

## American Resources Group

Signet Bank Building
374 Maple Avenue, East, Suite 210
Vienna, VA 22180
(703) 255-2700
*Keith A. Argow, President*

Provides monetary support for environmental services.

## Ashoka

1200 North Nash Street
Arlington, VA 22209
(202) 628-0370

Funds organizations/people who work for the public good in developing countries.

## Association of Conservation Engineers

c/o Alabama Department of
Conservation & Natural
Resources, Engineering Section
64 North Union Street
Montgomery, AL 36130
(205) 261-3476
*William Allinder,
Secretary/Treasurer*

## Atlantic Center for the Environment

39 South Main Street
Ipswich, MA 01938
(508) 356-0038
(508) 356-7322 FAX
*Lawrence B. Morris, Director*

Provides technical assistance to local groups, private and public.

## Bank Information Center

731 Eighth Street, SE
Washington, DC 20003
(202) 547-3800
*Chad Dobson, Secretary*

Works against environmentally damaging programs funded by the World Bank.

## BBN Laboratories, Inc.

10 Moulton Street
Cambridge, MA 02238
(617) 497-2559
*Donna M. Low, Manager of
Marketing and Communications*

Environmental acoustics specialists.

## Camp Fire Conservation Fund

230 Camp Fire Road
Chappaqua, NY 10514
(914) 941-2800
*George R. Lamb, President*

Funds activities of sportsmen and conservationists.

## Cascade Holistic Economic Consultants (CHEC)
P.O. Box 3479
Eugene, OR 97403
(503) 686-CHEC
*Randal O'Toole, Director/Forest Economist*

Consulting firm to forestry conservation groups.

## Charles A. Lindbergh Fund
708 S. Third Street, Suite 110
Minneapolis, MN 55415
(612) 338-1703
*Charles G. Houghton III, President*

Provides grants for activities which balance technological progress and environmental concerns.

## Chief Auto Parts
1515 Wade Drive
Seagoville, TX 75159
(214) 287-7474
*Michael H. Nanov, President and CEO*

Largest distributor of auto parts in America who sells an environmentally safe antifreeze that is non-toxic.

## Connecticut Fund for the Environment
Central Office, 152 Temple Street
New Haven, CT 06510
(203) 787-0646
*Suzanne Langille Mattei, Director*

Preserves and protects resources through legal action. Also publishes *Household Hazardous Waste*, a guide to disposal methods.

## Conservation Fund, The
1800 North Kent Street, Suite 1120
Arlington, VA 22209
(703) 525-6300
*Patrick F. Noonan, President*

Supports several programs and partnerships among other non-profit conservation organizations.

## Conservation International
1015 18th Street, NW, Suite 1000
Washington, DC 20036
(202) 429-5660
*Peter Seligmann, Chairman of the Board*
*Russell A. Mittermeier, President*

Funds conservation action in Latin America.

## Consumer Product Safety Commission
5401 Westbard Avenue
Bethesda, MD 20207
(301) 492-6500
*Anne Graham, Acting Chairman*

Responsible for product testing.

## Craighead Environmental Research Institute
Box 156
Moose, WY 83012
(307) 733-3387
*Frank C. Craighead, Jr., President*

Scientists studying cause and effect of man and his environment.

## E. F. Schumacher Society
Box 76, R.D. 3
Great Barrington, MA 01230
(413) 528-1737
*Robert Swann, President*

Promotes a decentralist philosophy, advocating small-scale technology based on philosopher E. F. Schumacher, author of *Small Is Beautiful.*

## Eco-home Network
4344 Russell Avenue
Los Angeles, CA 90027
(213) 662-5207
*Julia Russell, Founder*

Maintains a model home that's environmentally sound and publishes *Ecolution* newsletter.

**Economy Research Center**
502 South 19th Avenue
Bozeman, MT 59715
(406) 587-9591
*Monica Guenther, Spokesperson*

Environmental economics specialists.

**Environmental Opportunities**
P.O. Box 969
Stowe, VT 05672
(802) 253-9336
*Sanford Berry, Publisher*

Pamphlet lists environment-related job openings.

**Experimental Cities**
P.O. Box 731
Pacific Palisades, CA 90272
(213) 276-0686
*Genevieve Marcus, President*

A computer network linking non-profit groups. Also sponsors Earthlab research center.

**Genentech**
460 Pt. San Bruno Boulevard
South San Francisco, CA 94080
(415) 266-1000
*R. A. Swanson, CEO*

World's largest biotechnology company.

**Hambrecht & Quist**
1 Bush
San Francisco, CA 94104
(415) 576-3300

Operates the Environmental Technology Fund, a mutual fund which invests in companies developing new technologies.

**Heritage Trails Fund**
5301 Pine Hollow Road
Concord, CA 94521
(415) 672-5072
*George H. Cardinet, Executive Director*

Assists individuals and groups with fundraising to improve parks and multi-purpose trails.

**Industrial Designers Society of America**
Environmental Concern Committee
1142-E Walker Road
Great Falls, VA 22066
(703) 759-0100
*Budd Steinhilber, Committee Chairman*

Examines the environmental consequences of industrial design, such as in packaging and household goods.

**Institute of Environmental Sciences**
940 East Northwest Highway
Mt. Prospect, IL 60056
(312) 255-1561
*Janet A. Ehmann, Executive Director*

Studies the effect that rockets, missiles, satellites, ships, aircraft, ground vehicles, and nuclear radiation installations have on the natural environment.

**International Association of Natural Resource Pilots**
P.O. Box 309
Spooner, WI 54801
(715) 635-4169
*Thomas L. Hutchinson, President*

Seeks to promote increased knowledge and safety in natural resource aviation: in-flight observation of game and waterfowl and forest fire detection.

**International Center for the Solution of Environmental Problems.**
3818 Graustark Street
Houston, TX 77006
(713) 527-8711
*Joseph L. Goldman, Technical Director*

Science organization that attempts to anticipate upcoming environmental problems and provide solutions for avoiding or reducing them.

## Investor Responsibility Research Center
1755 Massachusetts Avenue, NW, Suite 600
Washington, DC 20036
(202) 939-6500
*Margaret Carroll, Executive Director*

Investigates how environmentally unsound products affect economic investments.

## Izaak Walton League of America Endowment
P.O. Box 824
Iowa City, IA 52244
(319) 351-7037
*Howard White, Honorary President*

Acquires unique natural areas for government agencies.

## J. N. (Ding) Darling Foundation
c/o Ralph Schlenker, Treasurer
P.O. Box 657
Des Moines, IA 50303
(515) 281-2371
*Christopher Koss, President, Board of Trustees and Chairman Executive Committee*

Gives educational grants for students in conservation and environmental studies.

## National Association of Conservation Districts
Conservation Technology Information Center
1220 Potter Drive, Room 170, Purdue Research Park
West Lafayette, IN 47906
(317) 494-5555
*John Becherer, Executive Director*

## National Center for Appropriate Technology
3040 Continental Drive, P.O. Box 3838
Butte, MT 59702
(406) 494-4572
*Charlie Green, Operations Vice-President*

Individuals and organizations who feel an ecological responsibility in their economic and financial activities.

## National Environmental Development Association
1440 New York Avenue, NW, Suite 300
Washington, DC 20005
(202) 638-1230
*Steven B. Hellem, Executive Vice-President*

Believes the financial burden of unjustified environmental measures could destroy industry's ability to provide the tax base for general social progress.

## National Environmental Technology Application Center
615 William Pitt Way
Pittsburgh, PA 15230
(412) 826-5511
*Lawrence T. McGeehan, Vice-President*

Studies and certifies the effectiveness and practicality of technology designed to aid the environment.

## National Park Foundation
P.O. Box 57473
Washington, DC 20037
(202) 785-4500
*Alan A. Rubin, President*

Chartered by U.S. Congress to provide private sector support of the National Park system.

**National Research Council, National Academy of Sciences, National Academy of Engineering, Institute of Medicine**
210 Constitution Avenue, NW
Washington, DC 20418
(202) 334-2000
*Frank Press, Chairman*

Independent adviser to the federal government.

**National Science Foundation**
1800 G Street, NW
Washington, DC 20550
(202) 357-9498
*Mary L. Good, Chairman, National Science Board*

Independent agency in the executive branch of the federal government that supports basic and applied research in the sciences.

**Nature Company**
750 Hearst Avenue, P.O. Box 2310
Berkeley, CA 94702
(800) 227-1114

Stores selling products that foster an appreciation for nature.

**New Alchemy Institute**
2376 Hatchville Road
East Falmouth, MA 02536
(617) 564-6301
*John Quinney, Executive Director*

Research and education for households and small farms on environmentally sound living and working.

**New Alternatives Fund**
295 Northern Boulevard
Great Neck, NY 11021
(516) 466-0808

Mutual fund which invests in environmentally oriented companies.

**Progressive Asset Management**
1814 Franklin
Oakland, CA 94612
(415) 834-3722
*Peter Camejo, President*

Mutual fund which invests in stocks of environmentally responsible waste management and landfill firms.

**Project Lighthawk**
P.O. Box 8163
Santa Fe, NM 87504
(505) 982-9656
*Michael M. Stewart, Executive Director*

Volunteers pilots to help in environmental projects.

**Real Goods Trading Company**
3041 Guidiville Road
Ukiah, CA 95482
(707) 468-9214
(800) 762-7325

A supplier of non-fossil fuel power equipment, energy-efficient lighting and many other environmentally progressive technologies.

**Resource Conservation Technology**
2633 North Calvert Street
Baltimore, MD 21218
(301) 366-1146

Sells a diverse collection of products for energy and water conservation, including water-saving toilets and radon mitigation products.

**Save the Planet Shareware**
Box 45
Pitkin, CO 81241
*Roger and Kathy Cox, Creators*

Devised an environmental computer database about the state of the earth.

## Seventh Generation

Department 60M89
10 Farrell Street
Burlington, VT 05403
(800) 456-1177
(800) 456-1139 FAX
*Alan Newman, President*

A mail-order business that specializes in solar-powered accessories and nontoxic cleaners.

## Sierra Club Foundation, The

730 Polk Street
San Francisco, CA 94109
(415) 776-2211
*Ronald P. Klein, President*

Finances the educational, literary, and scientific projects of groups working on the environment.

## Smithsonian Institution

Office of Fellowships and Grants
L'Enfant Plaza, SW, Room 7300
Washington, DC 20560
(202) 287-3271
*Roberta Rubinoff, Director*

Offers fellowships and grants to scientific projects concerned with many issues, including the environment.

## Texas Conservation Foundation

P.O. Box 12845, Capitol Station
Austin, TX 78711
(512) 463-2196
*John Hamilton, Executive Director*

Trustee for gifts to the state, including parklands.

## Virginia Outdoors Foundation

221 Governor Street
Richmond, VA 23219
(804) 786-5539
*Tyson B. Van Auken, Executive Director*

Coordinates private philanthropy for conservation and preservation.

## WEB

Ontario Environmental Network
456 Spadina Avenue, 2nd Floor
Toronto, Ontario M5T 2G8
Canada
(416) 929-0634

An electronic network of communication services and information retrieval for environmental groups.

## Wild Wings and Underhill Foundation

18 East 74th Street
New York, NY 10021
(212) 628-9077

Funds projects to protect and manage natural areas in developing countries.

## Wilderness Flyers

Seaplane Pilots Association
421 Aviation Way
Frederick, MD 21701
(301) 695-2082
*Robert A. Richardson, Executive Director*

Provides seaplane services to environmental groups.

## Working Assets Long Distance

230 California Street
San Francisco, CA 94111
(800) 522-7759
*Peter Barnes, Founder*

With partner US Sprint, this is a phone service that donates one percent of calling charges to environmental groups.

## World Bank

Information Center
2000 P Street, NW, #515
Washington, DC 20036
(202) 822-6630
*Barber Conable, President*

Funds environmental projects worldwide.

**World Future Society**
4916 South Elm Street
Bethesda, MD 20814
(301) 656-8274
*Edward S. Cornish, President*

Forecasts ideas about the future and publishes *The Futurist* magazine.

**World Nature Association**
P.O. Box 673
Silver Spring, MD 20901
(301) 593-2522
*Donald H. Messersmith, President*

Funds small conservation projects worldwide.

**Yaker Environmental Systems**
P.O. Box 18
Stanton, NJ 08885
(201) 735-7056

Develops computer software related to the environment.

# WATER

**Problem:** Rivers, lakes, oceans and groundwater have been contaminated with sewage and toxic pollutants dumped from industrial, agricultural, and urban sources. Our drinking water is in danger, as are the habitats of animals, particularly fish. At the same time, non-polluted water is no longer in unlimited supply.

**Solution:** Industry and agriculture must change production practices by recycling effluents, substituting non-toxic ingredients, or finding some other method of waste disposal. Everyone must become conscious that wasting good water, even at home, contributes to this serious problem.

This chapter includes government and non-government groups, international agencies, organizations, companies, and individuals mainly involved with: the pollution and protection of bays, canals, coastal areas, coral reefs, drinking water, lakes, oceans, rivers, and watersheds.

**Advisory Committee on Pollution of the Sea**
3 Endsleigh Street
London WC1H ODD England
388 2117
*Sir James Callaghan, President*

Promotes "the preservation of the seas of the world from pollution by human activities."

**Alpha Environmental**
P.O. Box 90218, 7748 Highway 290W, Suite 531
Austin, TX 78766
(512) 448-8927

Produces oil-eating microbes, a possible solution to oil spill cleanup.

**American Canal Society**
809 Rathton Road
York, PA 17403
(717) 843-4035
*William E. Trout, III, President*

Restores and preserves canals. Established 1972.

**American Clean Water Project**
107 Spyglass Lane
Fayetteville, NY 13066
(315) 637-4718
*Julia Portmore, President*
*Richard J. Fedele, Executive Director*

Established in 1982 to safeguard and restore drinking and swimming water.

**American Littoral Society**
Sandy Hook
Highlands, NJ 07732
(201) 291-0055
*Harold Nils Pelta, President*

National organization of professionals and amateurs for the conservation of coastal habitats, wetlands, estuaries.

**Coral Reef Conservation Center**
75 Virginia Beach Boulevard
Key Biscayne, FL 33149
(305) 361-4200
*Alexander Stone, Regional Director*

Florida Region
c/o Mote Marine Lab—City
Island
Sarasota, FL 33509
(813) 388-3301
*Susan Holderman, Regional
Director*

New England Region
P.O. Box 331
Woods Hole, MA 02543
(617) 548-2673
*William Sargent, Regional
Director*

New York Region
28 West 9th Road
Broad Channel, NY 11693
(718) 634-6467
*Donald Rieppe, Regional
Director*

## American Oceans Campaign
725 Arizona Avenue
Santa Monica, CA 90401
(213) 576-6162
*Bob Fulnick, Executive Director*

Organization for the West Coast conservation of oceans. Protesters against offshore drilling.

## American Rivers
801 Pennsylvania Avenue, SE,
Suite 373
Washington, DC 20003-2167
(202) 547-6900
*Kevin J. Coyle, President*

Supports strong state laws to preserve rivers and blocks unneeded dams. Established 1973.

## American Shore and Beach Preservation Association
P.O. Box 279
Middletown, CA 95461
(707) 987-2385
*Orville T. Magoon, President*

Conserves, develops, and restores beaches and shorefronts.

## American Society of Limnology and Oceanography
Virginia Institute of Marine
Science
College of William and Mary
Gloucester Point, VA 23062
(804) 642-7242
*Polly A. Penhale, Secretary*

Studies freshwaters and oceans.

## American Water Resources Association
5410 Grosvenor Lane, Suite 220
Bethesda, MD 20814
(301) 493-8600
*Peter E. Black, President*

Association of engineers and scientists to plan use of water resources.

## American Water Works Association
6666 West Quincy Avenue
Denver, CO 80235
(303) 794-7711
*John B. Mannion, Executive
Director*

Professionals involved with public water supplies.

## American Whitewater Affiliation
146 North Brockway
Palatine, IL 60067
(704) 483-5049
*Marge Cline, President*

Promotes river travel by various means of paddlecraft and also supports river conservation.

## Ametek Plymouth Products
502 Indiana Avenue
Sheboygan, WI 53081
(414) 457-9435

Produces activated-carbon water filters for home use.

## Arkansas Natural and Scenic Rivers Commission
P.O. Box 5670
North Little Rock, AR 72119
(501) 371-8134
*Basil Hicks, Jr., Chairman*

Prepares surveys and makes recommendations to the governor and the legislature.

## Association of State and Interstate Water Pollution Control Administrators
444 North Capitol Street, NW, Suite 330
Washington, DC 20002
(202) 624-7782
*Robbi J. Savage, Executive Director*

Promotes coordination among state agency programs and the EPA.

## Association of State Drinking Water Administrators
1911 North Fort Myer Drive, Suite 400
Arlington, VA 22209
(703) 524-2428
*G. Wade Miller, Executive Director*

Facilitates the exchange of information and experience among state drinking water program managers.

## Baltic Marine Environment Protection Committee
Helsinki Commission (HELCOM)
Mannerheimintie 12A
SF-00100 Helsinki 10, Finland
602 366
644 577 FAX

## Barrier Islands Coalition
122 East 42nd Street, Suite 4500
New York, NY 10168
(212) 949-0049
*Lisa Speer, Executive Director*

Seeks to preserve undeveloped coastal barrier islands and associated aquatic ecosystems.

## Bass Anglers for Clean Water, Inc.
P.O. Box 17900
Montgomery, AL 36141
(205) 272-9530
*Helen Swier, President*

Fisherfolk and freshwater advocates. Founded in 1970.

## California Department of Water Resources
P.O. Box 942836
Sacramento, CA 94236-0001
(916) 445-3157

## California Water Quality Control Board
Los Angeles Region
101 Centre Plaza Drive
Monterey Park, CA 91754
(818) 307-1282

## California Water Resources Control Board
901 P Street, P.O. Box 100
Sacramento, CA 95801-0100
(916) 445-3993
*W. Don Maughan, Chairman*

## Center for Health Action
P.O. Box 270
Forest Park Station
Springfield, MA 01108
(413) 782-2115
*Susan I. Pare, President*

Advocates the ban of water fluoridation.

## Center for Short-Lived Phenomena
P.O. Box 199, Harvard Square Station
Cambridge, MA 02238
(617) 492-3310
*Richard Golob, Director*

Believes oil spills will be a thing of the past if the Center's prevention methods are employed.

## Chesapeake Bay Foundation

162 Prince George Street
Annapolis, MD 21401
(301) 268-8816
*Godfrey A. Rockefeller, President*

Promotes the environmental welfare
and proper management of the bay.

## Clean Harbors Cooperative

P.O. Box 1375, 1200 State Street
Perth Amboy, NJ 08862
(201) 738-2438
*Edward M. Wirkowski, Manager*

A joint venture of 10 petroleum companies that make oil spill clean-up equipment and provide trained operators for
member companies.

## Clean Water Action Project

(also Clean Water Fund)
317 Pennsylvania Avenue, SE
Washington, DC 20003
(202) 547-1196
*David Zwick, Executive Director*

Citizen organization for national clean
water at an affordable cost.

## Clean Water Fund

Primary goals are to advance environmental and consumer protection with a
special focus on water pollution, toxic
hazards, solid waste management, and
natural resources and to develop the
grassroots strength of the environmental movement.

Chesapeake (MD, VA) Regional
Office
2500 North Charles Street
Baltimore, MD 21218
(301) 235-8808
*John Kable, Director*

Mid-Atlantic (DE, NJ, PA)
Regional Office
46 Bayard Street, Room 309
New Brunswick, NJ 08901
(201) 846-4224
*Ken Brown, Director*

Midwest (IA, MN, MI, ND,
WI) Regional Office
2395 University Avenue
St. Paul, MN 55114
(612) 645-0961
*Frank Hornstein, Co-director*

New England (MA, ME, NH,
RI, VT) Regional Office
186A South Street
Boston, MA 02111
(617) 423-4661
*Amy Goldsmith, Director*

South/Southwest (TX) Regional
Office
610 Brazos Street, Suite 101
Austin, TX 78701
(512) 474-0605
*David Zwick, Acting Director*

## Coastal Conservation Association

4801 Woodway, Suite 220 West
Houston, TX 77056
(713) 626-4222
*Walter W. Fondren III, Chairman
of the Board*

Concerned with marine, animal, and
plant life in coastal areas.

## Coastal Society

5410 Grosvenor Lane, Suite 110
Bethesda, MD 20814
(301) 897-8616
*Lauriston King, President*

Promotes the wise use of coastal environments.

## Columbia River Gorge Commission

P.O. Box 730
White Salmon, WA 98672
(509) 493-3323
*Stafford Hansell, Chairman*

Interstate (Washington and Oregon)
commission created to implement the
Columbia River Gorge National Scenic
Area Act.

## Connecticut River Watershed Council
125 Combs Road
Easthampton, MA 01027
(413) 584-0057
*Astrid T. Hanzalek, Chairman*

Four-state organization for the protection of surface and groundwater resources.

## Cousteau Society
930 West 21st Street
Norfolk, VA 23517
(804) 627-1144
*Jacques-Yves Cousteau, President and Chairman of the Board*

Educates the public through films and television on water conservation and appreciation. Also publishes *Calypso Log.*

California
8440 Santa Monica Boulevard
Los Angeles, CA 90069
(213) 656-4422

Europe
25 Wagram
F75017 Paris, France
14-766-02-46

New York
777 Third Avenue
New York, NY 10017
(212) 826-2940

## Ecological Water Products
1341 West Main Road
Middletown, RI 02840
(401) 849-4004

Sells water-efficient showerheads.

## Ecowater
P.O. Box 64420
St. Paul, MN 55164
(612) 739-5330

Sells activated-carbon water filters.

## Electrolux Water Systems
325 Riverside Avenue
Westport, CT 06880
(800) 356-8740

Sells high-volume, activated-carbon water purification systems.

## Freshwater Foundation
2500 Shadywood Road, Box 90
Navaree, MN 55392
(612) 471-8407
*Lindsay G. Arthur, Jr., Chairman*

Seeks understanding of water issues to encourage the proper use and management of surface water and groundwater.

## Friends of the River
Fort Mason Center, Building C
San Francisco, CA 94123
(415) 771-0400
*Larry Orman, Chairman*

The largest grassroots organization for river preservation in the U.S.

## Grand Calumet Task Force
2400 New York Avenue, Suite 605
Whiting, IN 46394

Attempting to clean up the polluted Grand Calumet River in Indiana and Illinois.

## Great Lakes Commission
The Argus II Building, 400 S. Fourth Street
Ann Arbor, MI 48103-4816
(313) 665-9135
*Henry G. Williams, Chairman*
*Michael J. Donahue, Executive Director*

Advisory agency for the eight Great Lakes states on water resources.

**Great Lakes United**
State University College at Buffalo
1300 Elmwood Avenue
Buffalo, NY 14222
(716) 886-0142
*John Jackson, President*

Coalition of organizations from eight states and two Canadian provinces conserving the Great Lakes-St. Lawrence River system.

Canada
P.O. Box 548, Station A
Windsor, Ontario N9A 6M6
Canada

Region I (Superior)
Upper Peninsula Environmental
  Coalition, P.O. Box 1014
Marquette, MI 49855
(906) 225-4323
*Scot Stewart, Regional Director*

Region II (Huron)
Saginaw Bay Advisory Council,
  1023 Brissette Beach
Kawkawlin, MI 48631
(517) 636-3141
*John Witzke, Regional Director*

Region III (Michigan)
Capitol Area Audubon, 13750
  Hardenburg Trail
Eagle, MI 48822
(517) 626-6680
*Joe Finkbeiner, Regional Director*

Region IV (Erie)
Conservation Department, 8000
  East Jefferson Avenue
Detroit, MI 48214
(313) 926-5269
*Pam Leisinger, Regional Director*

Region V (Ontario)
Canadian Environmental Law
  Association
243 Quent St. W., 4th Floor
Toronto, Ontario M5V 1Z4
Canada
(416) 977-2410
*Sarah Miller, Regional Director*

Region VI (St. Lawrence)
96 Grand Avenue
New York, NY 10013
(212) 226-2088
*Camilla Smith, Regional Director*

**Great Swamp Research Institute**
Office of the Associate Dean,
  College of Natural Sciences &
  Mathematics
Indiana University of Pennsylvania
305 Weyandt Hall
Indiana, PA 15705
(412) 357-2609
*Anne Harris Katz, President*
*Harvey Katz, Director*

Conducts research and education on swamps.

**Groundwater Newsletter**
Water Information Center
124 East Bethpage Road
Plainview, NY 11803
(516) 249-7634

Semi-monthly publication.

**Guadalupe-Blanco River Authority**
P.O. Box 271
Seguin, TX 78156-0271
(512) 379-5822
*John H. Specht, General Manager*

Protects and conserves water resources in a 10-county area.

**Heal the Bay**
1650A 10th Street
Santa Monica, CA 90404
(213) 399-1146
*Dorothy Green, President*

Dedicated to restoring Santa Monica Bay.

## Hudson River Sloop Clearwater
112 Market Street
Poughkeepsie, NY 12601
(914) 454-7673
*John Mylod, Executive Director*

Owners of the sloop Clearwater, who promote restoration and preservation of the Hudson River.

## IFO Water Management Products
2882 Love Creek Road
Avery, CA 95224
(209) 795-1758

Makes low-flush toilets

## Instituto de Estudios Amazonicos
Rua Itupava 1220
Curitiba, PR 80.040, Brazil
262-9494

Institute of Amazonian Studies.

## International Association on Water Pollution Research and Control
1 Queen Anne's Gate
London SW1H 9BT England

## International Bottled Water Association
113 North Henry Street
Alexandria, VA 22314
(703) 683-5213
*William Deal, Executive Vice-President*

Manufacturers, distributors and bottled water plants.

## International Committee on Coral Reefs
c/o Bernard Salvat
Ecole Pratique des Hautes Etudes,
Laboratoire de Biologie Marine
55 rue de Buffon
F-75005 Paris, France

## International Ocean Disposal Symposium
Department of Oceanography and Ocean Engineering
Florida Institute of Technology
Melbourne, FL 32901
(305) 768-8008
*Iver W. Duedall, Organizing Committee Chairman*

Researches the problems associated with disposing of wastes into the ocean.

## International Oceanographic Foundation
4600 Rickenbacker Causeway,
Virginia Key
Miami, FL 33149
(305) 361-4888
*Edward T. Foote II, President*

Concerned with the vital role of oceans in the world environment.

## International Rivers Network
301 Broadway, Suite B
San Francisco, CA 94133
(415) 986-4694

Dedicated to protecting the world's river systems.

## International Water Resources Association
205 North Mathews, University of Illinois
Urbana, IL 61801
(217) 333-6275
*Glenn E. Stout, Executive Director*

Worldwide organization of water managers, scientists, planners, educators, etc.

## International Water Supply Association
1 Queen Anne's Gate
London SW1H 9BT England
222 8111

Maintains a commission on pollution and protection of water supplies.

## Interstate Commission on the Potomac River Basin

6110 Executive Boulevard,
Suite 300
Rockville, MD 20852-3903
(301) 984-1908
*L. E. Zeni, Executive Director*

Member states: Maryland, Pennsylvania, Virginia, W. Virginia, and District of Columbia.

## Interstate Conference on Water Policy

955 L'Enfant Plaza, SW, 6th Floor
Washington, DC 20024
(202) 466-7287
*Carter Lee Kelly, Executive Director*

Facilitates interstate cooperation and exchange of information on all water quality and quantity matters.

## Lake Erie Cleanup Committee

3568 Brewster Road
Dearborn, MI 48120
(313) 271-8906
*John Chascsa, President*

Founded to stop the pollution of Lake Erie. Membership includes Ohio and Michigan citizen's groups.

## Lake Michigan Federation

59 East Van Buren, Suite 2215
Chicago, IL 60605
(312) 939-0838
*Henry T. Chandler, President of the Board of Directors*

Four-state coalition of citizens for the protection of Lake Michigan.

## League to Save Lake Tahoe

P.O. Box 10110
South Lake Tahoe, CA 95731
(916) 541-5388
*Graham B. Moody, President*

Attempts to preserve Lake Tahoe and surrounding area.

## Marine Coast Heritage Trust

167 Park Row
Brunswick, ME 04011
(207) 729-7366
*Harold E. Woodsum, Jr., Chairman*

Protects islands, coastlines and provides advice on open space protection.

## Maine Department of Marine Resources

State House, Station #21
Augusta, ME 04333
(207) 289-2291
*William J. Brennan, Commissioner*

## Marine Technology Society

1825 K Street, NW, Suite 203
Washington, DC 20006
(202) 775-5966
*Martin J. Finerty, Jr., General Manager*

Scientists, educators, engineers, and others with professional interests in the marine sciences.

## Mid-Atlantic Council of Watershed Associations

2955 Edge Hill Road
Huntingdon Valley, PA 19006
(215) 657-0830
*Bruce McNaught, President*

Advises new watershed associations.

## Minnesota-Wisconsin Boundary Area Commission

619 2nd Street
Hudson, WI 54106
(612) 436-7131
*James M. Harrison, Executive Director*

Planning for protection and use of the St. Croix and Mississippi rivers.

## Mono Lake Committee
P.O. Box 29
Lee Vining, CA 93541
(619) 647-6595
*Martha Davis, Executive Director*

Works toward preserving environmentally acceptable water levels by blocking diversions.

## National Association for State River Conservation Programs
P.O. Box 1467
Harrisburg, PA 17120
(717) 787-6816
*Roger Fickes, President*

Improves the efficiency and effectiveness of state river environmental preservation programs through communication and cooperation.

## National Boating Federation
1000 Thomas Jefferson Street,
   Suite 525
Washington, DC 20007
(202) 338-5718
*Ron Stone, Executive Director*

Protects interests of boaters.

## National Organization for River Sports
314 N. 20th Street, P.O. Box 6847
Colorado Springs, CO 80934
(719) 473-2466
*Gary Lacy, President*

Educates about white-water river sports and protects rights of access to river runners.

## National Water Alliance
1225 First Street, NW, Suite 300
Washington, DC 20005
(202) 646-0917 ext. 204
*Ron Linton, President*

Forms a consensus on water issues which is used by congressmen, senators, corporate leaders, environmental leaders, and academicians.

## National Water Center
P.O. Box 264
Eureka Springs, AZ 72632
(501) 253-9755
*Barbara Harmony, President*

Individuals interested in the maintenance and conservation of water resources.

## National Water Project
602 South King Street, Suite 402
Leesburg, VA 22075
(703) 771-8636
*Edwin L. Cobb, Executive Director*

Promotes adequate and affordable drinking water and wastewater disposal services to low-income rural areas.

## National Water Resources Association
955 L'Enfant Plaza North, SW,
   #1202
Washington, DC 20024
(202) 488-0610
*Raymond R. Rummonds, President*

Association to conserve and manage the water resources of 17 western states.

## National Water Supply Improvement Association
P.O. Box 102
St. Leonard, MD 20865
(301) 855-1173
*Jack C. Jorgensen, Executive Director*

Advances research and development programs in desalination, wastewater reclamation, and other water sciences.

## National Water Well Association
6375 Riverside Drive
Dublin, OH 43017
(614) 761-1711
*Richard Henkle, President*

Runs educational programs dealing with water topics.

**National Waterways Conference**
1130 17th Street, NW
Washington, DC 20036
(202) 296-4415
*Berdon Lawrence, Chairman*

Conducts water transportation programs dealing with their importance to the total environment.

**National Wetlands Technical Council**
1616 P Street, NW, Suite 200
Washington, DC 20036
(202) 328-5150
*Joseph S. Larson, Chairman*

Council of leading wetlands scientists.

**Nebraska Department of Water Resources**
State House Station, P.O. Box 94676
Lincoln, NE 68509
(402) 471-2363
*J. Michael Jess, Director*

**New England Association of Environmental Biologists**
25 Nashua Road
Bedford, NH 03102
(603) 472-5191
*Howard Davis, Information Officer*

Evaluates the effectiveness of environmental regulations protecting water quality.

**New England Interstate Water Pollution Control Commission**
85 Merrimac Street
Boston, MA 02114
(617) 367-8522
*David L. Clough, Chairman*

**New York Marine Sciences Research Center**
State University of New York
Stony Brook, NY 11794
(516) 632-8700
*J. R. Schubel, Director*

**New York St. Lawrence-Eastern Ontario Commission**
317 Washington Street
Watertown, NY 13601
(315) 785-2461
*Francis G. Healey, Co-chairman*
*William L. Curtis, Jr., Co-chairman*

Oversees development of the St. Lawrence region, with emphasis on conservation.

**North American Benthological Society**
c/o The Allen Press
1041 New Hampshire Street
Lawrence, KS 66044
(913) 864-4775
*Leonard C. Ferrington, President*

Researches aquatic ecology.

**North American Lake Management Society**
1000 Connecticut Avenue, NW, Suite 300
Washington, DC 20036
(202) 466-8550
*William Norris, President*

Promotes education and understanding of lakes, ponds, reservoirs, and their watersheds to advance their protection, restoration, and management.

**Ocean Magazine Associates**
2001 W. Main Street
Stamford, CT 06902
(203) 329-2787
*Richard Covington, Editor*

Publishes *Oceans* magazine.

**Oceanic Society**
1536 16th Street, NW
Washington, DC 20036
(202) 328-0098
*Clifton Curtis, President*

Dedicated to the protection and wise use of the oceans and marine environment.

## Ohio River Valley Water Sanitation Commission

49 East Fourth Street, Suite 815
Cincinnati, OH 45202
(513) 421-1151
*Alan H. Vicory, Jr., Executive Director and Chief Engineer*

Responsible for Ohio River water quality and pollution.

## Oregon Water Resources Department

3850 Portland Road, NE
Salem, OR 97310
(503) 378-3739
*William H. Young, Director*

## Passaic River Coalition

246 Madisonville Road
Basking Ridge, NJ 07920
(201) 766-7550
*Ella F. Filippone, Executive Administrator*

Seeks to resolve the problems of an urban river system, including pollution and waste disposal.

## Safe Water Coalition

150 Woodland Avenue
San Anselmo, CA 94960
(415) 453-0158
*John Lee, Researcher*

Research coalition created to stop fluoridation of water. Studies Gilbert Syndrome, a jaundice-like disease affected by fluoridation.

## Save San Francisco Bay Association

P.O. Box 925
Berkeley, CA 94701
(415) 849-3053
*Doris Sloan, President*

Dedicated to protecting the bay from additional pollution and water diversion.

## Save the Bay

434 Smith Street
Providence, RI 02908-3732
(401) 272-3540
*Christopher H. Little, President*

Largest citizens' environmental organization in Rhode Island, formed to protect Narragansett Bay.

## Scenic Hudson

9 Vassar Street
Poughkeepsie, NY 12601
(914) 473-4440
*Alexander E. Zagoreos, Chairman*

Preserves the Hudson River Valley.

## Sea Grant Consortium

287 Meeting Street
Charleston, SC 29401
(803) 727-2078
*Margaret A. Davidson, Director*

State agency that works with public research institutions to conserve coastal marine resources.

## Sea Grant Program

Virginia Graduate Marine Science
  Consortium
170 Rugby Road, Madison House
University of Virginia
Charlottesville, VA 22903
(804) 924-5965
*William L. Rickards, Director*

Sponsors research on marine fisheries and oceanography, public and commercial fishing industry outreach and education, and conferences on resource management and policy via college programs throughout the country.

## Seacoast Anti-Pollution League

Five Market Street
Portsmouth, NH 03801
(603) 431-5089
*Jane Doughty, Field Director*

Promotes the wise use of natural resources in the seacoast region of New England and seeks to prevent the Seabrook Nuclear Reactor from operating.

**Secretariat for the Protection of the Mediterranean Sea**
Placa Lesseps 1
E-08023 Barcelona, Spain
217 16 95

Countries in Africa, Asia, and Europe who facilitate information exchanges and action on protection of the Mediterranean.

**Sounds Conservancy**
P.O. Box 266, 43 Main Street
Essex, CT 06426
(203) 767-1933
*Christopher Percy, President*

Conserves, restores, and protects the marine region of Long Island, Fishers Island, Block Island, Rhode Island, Vineyard, and Nantucket sounds.

**South Carolina Water Resources Commission**
1201 Main Street, Suite 1100
Columbia, SC 29102
(803) 737-0800
*Erick B. Ficken, Chairman*

**Spill Control Association of America**
100 Renaissance Center, Suite 1575
Detroit, MI 48243
(313) 567-0500
*Marc K. Shaye, General Counsel*

Third-party contractors; manufacturers or suppliers of pollution control and containment equipment.

**Surfrider Foundation**
P.O. Box 2704, #86
Huntington Beach, CA 92647
(714) 846-3462
*Thomas P. Pratte, Executive Director*

Ocean-wave enthusiasts such as surfers interested in preserving ocean environments and coastal ecologies.

**Susquehanna River Basin Commission**
Interior Building
Washington, DC 20240
(202) 343-4091
*Robert J. Bielo, Executive Director*

Conserves and develops the river basin in cooperation with Maryland, New York, and Pennsylvania.

**Texas Water Development Board**
1700 North Congress
Austin, TX 78711
(512) 463-7847
*Walter W. Cardwell III, Chairman*
*Stuart S. Coleman, Vice-Chairman*

Handles statewide water resources planning and water development.

**Upper Colorado River Commission**
355 S. 4th East Street
Salt Lake City, UT 84111
(801) 531-1150
*Jack R. Ross, Chairman*

Interstate commission on water use.

**Upper Mississippi River Conservation Committee**
1830 Second Avenue
Rock Island, IL 61201
(309) 793-5800
*Bill Bertrand, Chairman*

Natural resource managers and biologists committed to the preservation and wise utilization of the Upper Mississippi River.

**Water Information Center**
125 East Bethpage Road
Plainview, NY 11803
(516) 249-7634

Publishes *Water Newsletter,* a bimonthly publication.

## Water Information Network

P.O. Box 4525
Albuquerque, NM 87106
(505) 262-1862

An organization linking grassroots water-quality groups in New Mexico.

## Water Pollution Control Federation

601 Wythe Street
Alexandria, VA 22314
(703) 684-2400
*Quincalee Brown, Executive Director*

Organization of state, regional, and foreign technical societies.

## Water Quality Association

4151 Naperville Road
Lisle, IL 60532
(312) 369-1600
*Peter Censky, Executive Director*

Manufacturers, distributors, and retailers of water treatment equipment and services.

## Water Resources Association of the Delaware River Basin

Box 867, Davis Road
Valley Forge, PA 19481
(215) 783-0634
*James A. Shissias, President*

Federation of businesses, industries, government agencies, citizens, and academicians who advocate the need for orderly conservation and development of water resources.

## Watertest

33 South Commercial Street
Manchester, NH 03101
(800) 426-8378

Mail-order water-testing laboratory.

## Western River Guides Association

7600 East Arapahoe Road
Englewood, CO 80112
(303) 771-0389
*Jerry Mallett, Executive Director*

Professional river outfitters who seek to preserve and protect western free-flowing rivers.

# WILDLIFE

**Problem:** Plant and animal species are facing extinction from air and water pollution, a growing human population, and diminishing land for habitats. The irretrievable loss of these species eliminates forever not only their potentially helpful and practical use to mankind, but also any future appreciation of them as fellow living creatures on this planet.

**Solution:** Safeguard animals and plants by protecting their habitats in the wild and developing ways to ensure their survival.

This chapter includes organizations, international agencies, government and non-government groups, and companies mainly concerned with: the protection and study of particular species of wild animals or wild animals in general, sport hunting, and fishing. (For more information on plant life, see *Land.*)

### African Wildlife Foundation
1717 Massachusetts Avenue, NW
Washington, DC 20036
(202) 265-8393
*Kermit Roosevelt, Honorary*
*Chairman*
*John H. Hemingway, Jr., Chairman*
*of the Board*

Foundation for the preservation of African wildlife.

### Alaska Department of Fish and Game
P.O. Box 3-2000
Juneau, AK 99802
(907) 465-4100
*Don W. Collingsworth,*
*Commissioner*

### Alberta Fish and Game Association
6924-104 Street
Edmonton, Alberta T6H 2L7
Canada
(403) 437-2342
*Jack Graham, President*

Established in 1908 to conserve and utilize the fish and wildlife of Alberta.

### Alberta Trappers Association
Bag 7000
Whitecourt, AB Edson AB
Alberta TOE 2LO, Canada
(403) 778-2602
*Jim Robison, President*

Founded in 1974 to help trappers and to provide courses on humane trapping.

### American Association of Wildlife Veterinarians
Department of Veterinary
Pathology
Iowa State University
Ames, IA 50011
(515) 294-3282
*Andrew Fix, Treasurer*

Veterinarians in state and federal wildlife resource agencies, agricultural agencies, and public health service agencies.

## American Association of Zoo Veterinarians
Philadelphia Zoo
34 Street and Girard Avenue
Philadelphia, PA 19104
(215) 387-9094
*Wilbur Armand, Executive Director*

Veterinarians in zoo and wildlife medicine.

## American Association of Zookeepers
National Headquarters
635 Gage Boulevard
Topeka, KS 66606
(913) 272-5821
*Oliver Claffey, President*

Advocates quality animal care, animal keeping as a profession. Established 1967.

## American Association of Zoological Parks and Aquariums
Ogleby Park
Wheeling, WV 26003
(304) 242-2160
*David G. Zucconi, President*

Publishes *Zoological Parks and Aquariums in the Americas* directory.

## American Bass Association
886 Trotters Trail
Wetumpka, AL 36092
(205) 567-6035
*Bob Parker, President*

Promotes bass fishing, protecting fishery reserves, and "appreciation for life giving waters."

## American Birding Association
P.O. Box 6599
Colorado Springs, CO 80934
(800) 634-7736
*Allan R. Keith, President*

Studies birds in their natural environment. Publishes *Birding* and *Winging It.*

## American Cetacean Society
P.O. Box 2639
San Pedro, CA 90731-0943
(213) 518-6279
(213) 548-6950 FAX
*Susan Lafferty, President*

Conservation and education on the protection of marine mammals: whales, dolphins, and porpoises. Established 1967.

## American Fisheries Society
5410 Grosvenor Lane
Bethesda, MD 20814
(301) 897-8616
*Larry A. Nielsen, President*

Promotes fish conservation and wise utilization of resources. Established 1870.

Fish Health Section, New York
DEC, Fish Disease Control
Unit
83214 Fish Hatchery Road
Rome, NY 13440
(315) 337-0910
*John Schachte, President*

Marine Fisheries
Section—University of Miami
Cimas/RSMAS—Rickenbacker
Causeway
Miami, FL 33149
(305) 361-4185
*Dr. William W. Fox, Jr.,
President*

NMFS, NW, and Alaska
Fisheries Center
7688 Sand Point Way, NE
Seattle, WA 98115
(206) 526-4294
*Vidar G. Wespestad, President*

North Central Division—
Missouri Department of
Conservation
2901 West Truman Boulevard
Jefferson City, MO 65102
(314) 751-4115
*Lee Redmond, President*

Northeastern Division—Maine Cooperative Fish and Wildlife Research Unit
University of Maine, Department of Zoology
Orono, ME 04469
(207) 581-2582
*Dr. John Moring, President*

Socioeconomics Section-Sport Fishing Institute
1010 Massachusetts Avenue, NW, Suite 100
Washington, DC 20004
(202) 898-0770
*David B. Rockland, President*

Southern Division—Department of Zoology, North Carolina State University, Campus Box 7617
Raleigh, NC 27695
(919) 737-2741
*Dr. Richard L. Noble, President*

Water Quality Section—U.S. Fish and Wildlife Service
NFRC-Field Research Station, P.O. Box 936
La Crosse, WI 54602
(608) 783-6451
*James G. Wiener, President*

Western Division
10685 Jackson Road
Sacramento, CA 95830
(916) 362-8373
*Cay Goude, President*

## American Horse Protection Association
1000 29th Street, NW, Suite T-100
Washington, DC 20007
(202) 965-0500
*William L. Blue, President and Chairman of the Board*

Concerned with the welfare of wild and domestic horses.

## American Ornithologists' Union
National Museum of Natural History
Smithsonian Institution
Washington, DC 20560
(202) 357-1970
*Glen E. Woolfenden, President*

Organization of bird specialists.

## American Society of Ichthyologists and Herpetologists
Section of Ecology and Systematics
Corson Hall, Cornell University
Ithaca, NY 14853-2701
(607) 255-6582
*F. Harvey Pough, President*

Studies fish and amphibious reptiles.

## American Society of Mammalogists
501 Widtsoe Building, Dept. of Zoology
Brigham Young University
Provo, UT 84602
(801) 378-2492
*H. Duane Smith, Secretary-Treasurer*

Association for scientists who study mammals.

## American Society of Zoologists
104 Sirius Circle
Thousand Oaks, CA 91360
(805) 492-3585
*Albert F. Bennett, President*

Association of professional zoologists. Founded 1890.

## American Veterinary Medical Association
930 North Meacham Road
Schaumburg, IL 60196
(312) 605-8070
*Arthur Freeman, Executive Vice-President*

## Animal Welfare Institute

P.O. Box 3650
Washington, DC 20007
(202) 337-2333
*Christine Stevens, President*

Goal is to improve conditions for animals. Also includes the Save the Whales Campaign. Founded 1951.

## Animals Agenda Magazine

456 Monroe Turnpike, P.O. Box 456
Monroe, CT 06468
(203) 452-0446
*Kim Bartlett, Editor*

America's first publication devoted to the protection of animals, wild and domestic.

## Arctic International Wildlife Range Society

c/o Nancy R. Leblond
917 Leovista Avenue
North Vancouver, British Columbia V1R 1R1 Canada
(604) 986-0586

Works to promote an international wildlife reserve along the border between Alaska and Canada.

## Arizona Game and Fish Department

2222 West Greenway Road
Phoenix, AZ 85023
(602) 942-3000
*Duane L. Shroufe, Director*

## Arkansas Game and Fish Commission

#2 Natural Resources Drive
Little Rock, AR 72205
(501) 223-6300
*Steve N. Wilson, Director*

## Artek

P.O. Box 145
Antrim, NH 03440
(603) 588-6825

Produces "ivory alternative" for use in scrimshaw.

## Association for Fish and Wildlife Enforcement Training

Alberta Fish and Wildlife Department
Main Floor, N. Tower, Petroleum Plaza—9945 108 Street
Edmonton, Alberta T5K 2G6 Canada
(403) 427-6735
*Rod Hasay, President*

Association of U.S. and Canadian enforcement officers.

## Association of Field Ornithologists

Division of Birds—National Museum of Natural History
Smithsonian Institution
Washington, DC 20560
(202) 357-2031
*Peter Cannell, President*

## Association of Midwest Fish and Wildlife Agencies

Michigan Department of Natural Resources
P.O. Box 30028
Lansing, MI 48909
(517) 373-1263
*David Hales, President*

## Atlantic Salmon Federation

International Headquarters
P.O. Box 429
St. Andrews, New Brunswick EOG 2XO Canada
(506) 529-8889
(506) 529-4438 FAX
*David R. Clark, President*

**Atlantic States Marine Fisheries Commission**
1400 Sixteenth Street, NW
Washington, DC 20036
(202) 387-5330
*Irwin M. Alperin, Executive Director*

Organization of 15 Eastern Seaboard states for the promotion and protection of fisheries.

**Audubon Naturalist Society of the Central Atlantic States**
8940 Jones Mill Road
Chevy Chase, MD 20815
(301) 652-9188
*Anthony White, President*

Independent Audubon organization since 1896. Operates Woodend Wildlife Sanctuary.

**Audubon Society**
Sociedad Audubon de Panama
Apartado 2026
Balboa, Panama

**Audubon Society of Rhode Island**
12 Sanderson Road
Smithfield, RI 02911
(401) 231-6444
*Irving M. Leven, President*

**Bass Anglers Sportsman Society**
P.O. Box 17900, One Bell Road
Montgomery, AL 36141
(205) 272-9530
*Helen Sevier, Chairman-CEO*

Works to prevent the pollution of fishing waters.

**Bass Research Foundation**
1001 Market Street
Chattanooga, TN 37402
(615) 756-2514
*H. William Bucher, Executive Director*

Promotes scientific bass management.

**Bat Conservation International**
P.O. Box 162603
Austin, TX 78716
(512) 327-9721
*Verne R. Read, Chairman*

Promotes the protection of bats and their usefulness to the environment.

**Bateman, Robert McLellan**
c/o Center Court
Venice, FL 34292

Painter of animal subjects.

**Beaver Defenders**
Unexpected Wildlife Refuge
P.O. Box 765
Newfield, NJ 08344
(609) 697-3541
*William Bey, President*

Preserves and protects wildlife, especially beavers.

**Billfish Foundation**
2051 NW 11th Street
Miami, FL 33125
(305) 649-8930
(305) 649-7842 FAX
*Winthrop P. Rockefeller, President and Chairman*

Dedicated to the preservation of the billfish worldwide.

**Boone and Crockett Club**
241 South Fraley Boulevard
Dumfries, VA 22026
(703) 221-1888
*James H. Duke, Jr., President*

Conserves wildlife and does comprehensive surveys of big game species. Founded 1887.

**Boone and Crockett Club Foundation**
225 11th Avenue, Suite 21
Helena, MT 59601
(406) 442-6350
*William L. Searle, President*

Owns and operates the Theodore Roosevelt Memorial Ranch, a working cattle ranch which serves as a wintering area for wildlife and a wildlife research facility.

## Bounty Information Service
Stephens College Post Office
Columbia, MO 65215
(314) 876-7186
*H. Charles Lavin, Director and Editor*

Publishes *Bounty News.* Founded in 1966 to promote the removal of bounties from wild animals in North America.

## British Columbia Waterfowl Society
5191 Robertson Road
Delta, British Columbia V4K 3N2
Canada
(604) 946-6980
*Craig Runyan, President*

Operates the George C. Reifel Migratory Bird Sanctuary.

## British Columbia Wildlife Federation
5659 176th Street
Surrey, British Columbia V3S 4C5
Canada
(604) 576-8288
*Steve Head, President*

Organization formed to promote the wise use of the environment in British Columbia.

## Brookfield Zoo
Brookfield, IL 60512
(708) 485-0263
*George Rabb, Director*

One of America's largest and most important zoos.

## Brooks Bird Club
707 Warwood Avenue
Wheeling, WV 26003
(304) 547-5253
*Nevada Laitsch, President*

Named after A. B. Brooks, naturalist. Founded in 1932.

## Brooks Nature Center
Oglebay Institute
Oglebay Park
Wheeling, WV 26003
(304) 242-6855
*Sue B. Stroyls, Director, Nature Education*

## Brotherhood of the Jungle Cock
P.O. Box 576
Glen Burnie, MD 21061
(301) 761-7727
*Bill Hampt, President*

Teaches youth the true meaning of conservation of game fishes.

## California Department of Fish and Game
1416 Ninth Street
Sacramento, CA 95814
(916) 445-3531
*Peter F. Bontadelli, Director*

## California Wildlife Defenders
P.O. Box 2025
Hollywood, CA 90078
(213) 663-1856
*Lila Brooks, Director*

Promotes eradication of the prejudice against predators.

## Canadian Nature Federation
453 Sussex Drive
Ottawa, Ontario K1N 6Z4 Canada
(613) 238-6154
*Jacques Prescott, President*

Naturalists' umbrella organization.

## Canadian Society of Environmental Biologists
P.O. Box 962, Station F
Toronto, Ontario M4Y 2N9
Canada

## Canadian Wildlife Federation
1673 Carling Avenue
Ottawa, Ontario K2A 3Z1 Canada
(613) 725-2191

One of Canada's largest environmental groups, founded in 1961.

Alberta
6016-105 Street
Edmonton, Alberta T6H 2N4
Canada
(403) 434-1198
*John Graham, Director*

British Columbia
204 1560 Hillside Avenue
Victoria, British Columbia
V8T 5B8 Canada
(604) 592-6610
*Stewart Reeder, Director*

Manitoba
P.O. Box 387
Carberry, Manitoba ROK OHO
Canada
(204) 834-3130
*Robert Barton, Director*

New Brunswick
887 Union Street
Fredericton, New Brunswick
E3A 3P7 Canada
(506) 450-8182
*Dale Stickles, Director*

Newfoundland and Labrador
14 Fairhaven Place
St. John's, Newfoundland
A1E 4S1 Canada
(709) 737-7263
*Richard Bouzan, Director*

Northwest Territories
P.O. Box 1636
Hay River, Northwest
Territories X0E 0R0 Canada
(403) 874-2627
*Sallyann Herbert, Director*

Nova Scotia
R.R. No. 2
Tantallon, Nova Scotia B0J 3J0
Canada
(902) 826-7507
*Paul Crawford, Director*

Ontario
3 Thunder Drive
Dryden, Ontario P8N 1V9
Canada
(807) 223-6336
*Charles Alexander, Director*

Prince Edward Island
Suite 9, Box 91, R.R. No. 7
Charlottetown, Prince Edward
Island C1A 7J9 Canada
(902) 566-0387
*Pat Doyle, Director*

Quebec
319 Est. rue St.-Zotique
Montreal, Quebec H2S 1L5
Canada
(514) 271-2487
*Andre Pelletier, Director*

Saskatchewan
R.R. No. 1
Tisdale, Saskatchewan S0E 1T0
Canada
(306) 873-5472
*Derrek Stanley, Director*

Yukon Territory
9 Klondike Road
Whitemore, Yukon Territory
Y1A 3L8 Canada
(403) 667-5716
*Doug Phillips, Director*

## Canvasback Society, The
P.O. Box 101
Gates Mills, OH 44040
(216) 443-2340
*Oakley V. Andrews, President*

Works for conservation of the canvasback duck of North America.

## Caribbean Conservation Association
Savannah Lodge, The Garrison
St. Michael, Barbados
(809) 426-5373
*Yves Renard, President*

Associate of the National Wildlife Federation.

## Caribbean Conservation Corporation

P.O. Box 2866
Gainesville, FL 32602
(904) 373-6461
*Charles D. Webster, Chairman of the Board*

Maintains research station at Tortuguero, Costa Rica, for marine turtles.

## Center for Marine Conservation

1725 DeSales Street, NW,
Suite 500
Washington, DC 20036
(202) 429-5609
*William Y. Brown, Chairman of the Board*

Conservation of endangered marine species and their habitats.

## Chelonia Institute

P.O. Box 9174
Arlington, VA 22209
(703) 524-4900
*Robert W. Truland, Director*

Focuses on the conservation of marine turtles.

## Colorado River Fish and Wildlife Council

241 North Vine Street, E, #401
Salt Lake City, UT 84103-1962
(602) 942-3000
*Duane Shroufe, Chairman*

Association of wildlife organizations from seven states.

## Columbia Basin Fish and Wildlife Authority

2000 SW First Avenue, Suite 170
Portland, OR 97201-5346
(503) 326-7031
*John Smith, Chairman*

Implements Pacific Northwest Electric Power Planning and Conservation Act. Consists of two federal agencies, five state agencies, and 13 Indian tribes.

## Committee for Conservation and Care of Chimpanzees

3819 48th Street
Washington, DC 20016
(202) 362-1993
*Geza Teleki, Chairman*

Representatives in 25 countries concerned with the survival of wild and captive chimpanzees.

## Committee for the Preservation of the Tule Elk

P.O. Box 3696
San Diego, CA 92103
(619) 485-0626
*Jolene W. Steigerwalt, Secretary*

Goal is to protect and preserve all wildlife, particularly rare species such as the Tule Elk.

## Committee to Abolish Sport Hunting

Box 43
White Plains, NY 10605
(914) 428-7523
*Luke A. Dommer, Chairman*

Runs a 60-acre refuge in the Catskill Mountains and patrols other wildlife refuge areas.

## Cooper Ornithological Society

Department of Biology
University of California
Los Angeles, CA 90024-1606
(602) 523-4307
*Russell P. Balda, President*

Aids conservation, observation, and cooperative study of birds.

## Defenders of Wildlife

1244 19th Street, NW
Washington, DC 20036
(202) 659-9510
*M. Rupert Cutler, President*

Founded in 1947 to preserve and protect wildlife ecosystems.

California-Nevada
5604 Rosedale Way
Sacramento, CA 95822
(916) 442-6386
*Richard Spotts, Regional
Representative*

Northern Rockies
1534 Mansfield Avenue
Missoula, MT 59801
(406) 549-0761
*Hank Fischer, Regional
Representative*

Southwest
13795 North Como Drive
Tucson, AZ 85741
(602) 297-1434
*Aubrey Stephen Johnson,
Regional Representative*

## Desert Bighorn Council
P.O. Box 5431
Riverside, CA 92517
(714) 351-6394
*Donald J. Armentrout,
Secretary-Treasurer*

Conserves the desert bighorn sheep in
Mexico and the United States.

## Desert Fishes Council
407 West Line Street
Bishop, CA 93514
(619) 872-1171
*Edwin P. Pister, Executive
Secretary*

Protects ecosystems of North American deserts.

## Desert Tortoise Council
5319 Cerritos Avenue
Long Beach, CA 90805
(213) 422-6172
*Dan Pearson, Co-chairman*
*Mike Gusti, Co-chairman*

Protects *Gopherus agassizi*, the desert
tortoise.

## Dolphin Research Center
P.O. Box 2875
Marathon Shores, FL 33052
(305) 289-0002
*Jayne Rodriguez, President*

Conducts research on dolphin lan-
guage, their learning ability, and biology.

## Ducks Unlimited
One Waterfowl Way
Long Grove, IL 60047
(708) 438-4300
*Harry D. Knight, President*

Founded in 1937 to preserve waterfowl.

Canada
1190 Waverley Street
Winnipeg, Manitoba R3T 2E2
Canada
(204) 477-1760
*Arthur L. Irving, Chairman of
the Board*

Mexico
Apartado 776, 64000 Monterrey
Nuevo Leon, Mexico
335-4032

## Eagle Foundation, The
300 East Hickory Street
Apple River, IL 61001
(815) 594-2259
*Terrence N. Ingram, President and
Executive Director*

Preserves vital eagle habitats.

## Elephant Interest Group
Department of Biological Sciences
Wayne State University
Detroit, MI 48202
(313) 540-3947
*Hezy Shoshani, Contact*

Works to promote public knowledge
and preservation of elephants.

## Elsa Wild Animal Appeal

P.O. Box 4572
North Hollywood, CA 91617
(818) 761-8387
*A. Peter Rasmussen, Jr., General
Manager*

International organization dedicated to the preservation of wildlife. Founded by Joy Adamson, author of *Born Free*.

Illinois
994 South Saylor Avenue
Elmhurst, IL 60126
*Donald A. Rolla, Regional
Representative*

Louisiana
1540 Chateau Circle
Lake Charles, LA 70605
*Laura Lanza, Regional
Representative*

New Hampshire
Laconia, NH 03246
*Pamela Clark, Regional
Representative*

## Endangered Species Act Reauthorization Coordinating Committee

1725 De Sales Street, NW,
Suite 500
Washington, DC 20036
(202) 429-5609
*Roger E. McManus, Vice-President*

Works to extend the Endangered Species Act and to ensure its effective implementation.

## European Association for Aquatic Mammals

358 avenue Mozart
F-06600 Antibes, France

## European Cetacean Society

c/o Zoology Department, Oxford
University
South Parks Road
Oxford OX1 4PS England

Studies and protects whales, porpoises, etc.

## European Committee for the Protection of Fur Animals

16 rue de Vriere
B-1020 Brussels, Belgium

## Fauna and Flora Preservation Society

79-83 North Street
Brighton BN1 1ZA England
820 445

Major wildlife preservation society. Founded in 1903.

## Federation of Fly Fishers

P.O. Box 1088
West Yellowstone, MT 59758
(406) 646-9541
*Keith Groty, President*

Promotes fly fishing and fish conservation.

## Federation of Ontario Naturalists

355 Lesmill Road
Don Mills, Ontario M3B 2W8
Canada
(416) 444-8419

One of Canada's oldest environmental groups, founded in 1931.

## Felicidades Wildlife Foundation

Box 490
Waynesville, NC 28786
(704) 926-0192
*George R. Collett, Jr., President*

Conducts wildlife rehabilitation workshops, slide lectures, and programs for children.

## Fish and Wildlife Reference Service

5430 Grosvenor Lane
Bethesda, MD 20814
(301) 492-6403
*Mary J. Nickum, Project Leader*

Information retrieval system and clearinghouse for research reports.

**Florida Game and Fresh Water Fish Commission**
620 South Meridian Street
Tallahassee, FL 32399-1600
(904) 388-1960
*Robert M. Brantly, Executive Director*

**Foundation for North American Wild Sheep**
720 Allen Avenue
Cody, WY 82414
(307) 527-6261
*Karen Werbelow, Office Manager*

Conducts wildlife studies, improves habitats, and promotes game management of wild sheep.

**Foundation for the Preservation and Protection of the Przewalski Horse**
380 Animal Sciences Building, VPI & SU
Blacksburg, VA 24061
(703) 961-5252
*Arden N. Huff, Educational Director*

Preserves the only wild horse species that exists in the wild in Europe and Asia.

**Friends of Africa in America**
330 South Broadway
Tarrytown, NY 10591
(914) 631-5168
*Clement E. Merowit, President*

Supports wildlife programs, primarily in East Africa.

**Friends of Animals and Their Environment**
P.O. Box 7283
Minneapolis, MN 55407
(612) 690-4997
*Howard Goldman, Coordinator*

Promotes the ban of the leghold trap and other trapping abuses.

**Friends of the Sea Lion Marine Mammal Center**
20612 Laguna Canyon Road
Laguna Beach, CA 92651
(714) 494-3050
*W. H. Ford, Executive Officer*

Aids sick and injured marine animals and returns them to their natural habitats.

**Friends of the Sea Otter**
P.O. Box 221220
Carmel, CA 93922
(408) 625-6965
*Margaret W. Owings, President*

Protects and maintains the endangered sea otter along the California coast.

**Funds for Animals**
200 West 57th Street
New York, NY 10019
(212) 246-2096
*Cleveland Amory, President*

Strives to save endangered species, preserve wildlife, and promote humane treatment.

Animal Trust Sanctuary
18740 Highland Valley Road
Ramona, CA 92065
*Chuck Traisi, Manager*

Black Beauty Ranch
P.O. Box 367
Murchison, TX 75778
*Bill Saxon, Manager*

Chicago
642 West Buckingham Place
Chicago, IL 60657
(312) 943-6700
*Margaret Asproyerakas, Field Officer*

Denver
1291 Gaylord
Denver, CO 80219
(303) 333-8294
*Sherri Tippie, Field Officer*

Florida
4069 Coquina Drive
Sanibel Island, FL 33954
(813) 472-2825
*George Campbell, Field Officer*

Florida
8125 Bryant Road
Lakeland, FL 33809
(813) 984-1157
*Donna Gregory, Field Officer*

Louisiana
P.O. Box 10676
Jefferson, LA 70181
(803) 834-5010
*Sid Rosenthal, Field Officer*

Michigan
2841 Colony Road
Ann Arbor, MI 48104
(313) 971-4632
*Doris Dixon, Field Officer*

Minnesota
P.O. Box 427
Spring Park, MN 55384
(612) 471-9305
*Barbara Zell, Field Officer*

New York
98 Woodlawn Avenue
Albany, NY 12208
(518) 438-6369
*Dorit Stark-Riemer, Field Officer*

Ohio
4132 West 214 Street
Fairview Park, OH 44126
(216) 331-9336
*Gregory Gorney, Field Officer*

Pennsylvania
21 Stoney Hill Road
New Hope, PA 18938
(215) 862-5031
*Cynthia Branigan, Field Officer*

San Francisco
Fort Mason Center, Bldg. 310
San Francisco, CA 94123
(415) 474-4020
*Virginia Handley, Field Officer*

South Carolina
Rt. 2, Box 559
Simponsville, SC 29681
(803) 963-4389
*Caroline Gilbert, Field Officer*

Texas
10022 Cedar Creek
Houston, TX 77001
(713) 952-5024
*Dana Forbes, Field Officer*

Toronto
2335 Lakeshore Boulevard, Apt. 715
Toronto, Ontario M8V 1B9
Canada
(416) 251-4571
*Marlene Lakin, Field Officer*

Washington, DC
1129 20th Street, NW, Suite 500
Washington, DC 20036
(202) 293-0150
*Barry Schochet, Field Officer*

## Fundacion Charles Darwin para las Islas Galapagos

Casilla 3991
Quito, Ecuador
527-912
*Craig MacFarland, President*

Charles Darwin Foundation for the Galapagos Islands. Established in 1959.

## Fur Takers of America

Rt. 3, Box 211A1
Aurora, IN 47001
(812) 926-1049
*Marsha Walston, Treasurer*

Fur trappers, buyers, trapping suppliers, hunters, fur dressers, and conservationists who seek to promote humane trapping.

## Future Fisherman Foundation

1 Berkeley Drive
Spirit Lake, IA 51360
(800) 237-5539
*Michael Fine, Executive Director*

Promotes participation and education in fishing and enhancing the environment.

## Game Conservation International

P.O. Box 17444
San Antonio, TX 78217
(512) 824-7509
*Harry L. Tennison, President*

Organization of hunter conservationists which promotes anti-poaching programs.

## George Miksch Sutton Avian Research Center

P.O. Box 2007
Bartlesville, OK 74005
(918) 336-7778
*John A. Brock, Board Chairman*

Strives for protection of all birds, particularly the American bald eagle and the Andean condor.

## Golden Lion Tamarin Management Committee

Department of Zoological Research
National Zoological Park
Washington, DC 20008
(202) 673-4825
*Devra Kleiman, Executive Officer*

Fosters conservation of the Golden Lion tamarin, a small marmoset indigenous to South America.

## Goodall, Jane

P.O. Box 26846
Tucson, AZ 85726

Chimpanzee expert.

## Gopher Tortoise Council

Florida Museum of Natural
    History
University of Florida
Gainesville, FL 32611
(904) 392-1721
*Ray Ashton, Co-chairman*
*George Heinrich, Co-chairman*

Protects the survival of *Gopherus polyphemus*, the gopher tortoise.

## Gorilla Foundation

P.O. Box 620-530
Woodside, CA 94062
(415) 851-8505
*Ronald H. Cohn, Secretary*

Promotes conservation, propagation, and behavioral study of apes, especially gorillas.

## Great Bear Foundation

P.O. Box 2699
Missoula, MT 59806
(406) 721-3009
*Lance Olsen, President*

Concerned with the conservation of bears, particularly grizzly bears. Also publishes *Bear News*.

## Great Lakes Fishery Commission

1451 Green Road
Ann Arbor, MI 48105
(313) 662-3209
*Carlos M. Fetterolf, Jr., Executive Secretary*

Canada-U.S. cooperative commission.

## Great Lakes Sport Fishing Council

293 Berteau
Elmhurst, IL 60126
(312) 941-1351
*Dan Thomas, President*

Clearinghouse for information about the future of sport fishing.

## Greenpeace

London
6 Endsleigh Street
London WC1H ODX England
387 5370

Activist organization particularly concerned with wildlife. Not connected to Greenpeace International.

## Gulf and Caribbean Fisheries Institute

c/o South Carolina Sea Grant
Consortium
287 Meeting Street
Charleston, SC 29401
(803) 727-2078
*Melvin Goodwin, Executive
Secretary*

## Gulf States Marine Fisheries Commission

P.O. Box 726
Ocean Springs, MS 39564
(601) 875-5912
*Larry B. Simpson, Executive
Director*
*Taylor F. Harper, Commission
Chairman*

Interstate association comprised of Alabama, Florida, Louisiana, Mississippi, and Texas to promote proper utilization of fisheries on the Gulf Coast.

## Hai-Bar Society for the Establishment of Biblical National Wildlife Reserves in Israel

c/o Nature Reserves Authority
78 Yirmeyahu Street
Jerusalem 94467, Israel

Works to establish reserves for animals referred to in the Bible.

## Hasti Friends of the Elephants

P.O. Box 477
Petaluma, CA 94953
(707) 878-2395

Works to save both Asian and African elephants from extinction.

## Hawk Migration Association of North America

377 Loomis Street
Southwick, MA 01077
(413) 569-3335
*Seth Kellogg, Chairman*

Studies birds of prey, particularly migration patterns.

## Hawk Mountain Sanctuary Association

Rt. 2, Box 191
Kempton, PA 19529
(215) 756-6961
*Stanley E. Senner, Executive
Director*

Maintains 2,200-acre Hawk Mountain Sanctuary.

## Hawk Trust, The

c/o Birds of Prey Section
London Zoo, Regents Park
London NW1 4RY England
*Jane Fenton, Chairman*

Protects birds of prey.

## Humane Society of the United States

2100 L Street, NW
Washington, DC 20037
(202) 452-1100
*K. William Wiseman, Chairman of
the Board*

Protects domestic and wild animals.

## Idaho Fish and Game Department

600 South Walnut, Box 25
Boise, ID 83707
(208) 334-3700
*Jerry M. Conley, Director*

## Inland Bird Banding Association

R.D. 2, Box 26
Wisner, NE 68791
(402) 529-6679
*John J. Flora, President*

Coordinates bird banding activities with government officials and private organizations.

**Inter-American Tropical Tuna Commission**
c/o Scripps Institute of
 Oceanography
La Jolla, CA 92093
(619) 546-7100
*James Joseph, Director*

Member nations: U.S., France, Japan, Nicaragua, Panama. Investigates and conserves tuna and dolphin resources.

**International Association for Bear Research and Management**
c/o ADF&G 333 Raspberry Road
Anchorage, AK 99518-1599
(602) 567-5111
*Al LeCount, President*

Sponsors International Conference on Bears and cooperation among agencies and universities.

**International Association of Fish and Wildlife Agencies**
444 North Capitol Street, NW,
 Suite 534
Washington, DC 20001
(202) 624-7890
*William A. Molini, President*

Wildlife association encompassing governments in the Western hemisphere.

**International Bird Rescue Research Center**
699 Potter Street
Berkeley, CA 94710
(415) 841-9086
*Alice B. Berkner, Executive
 Director*

Rehabilitates water birds oiled by slicks and develops new techniques for their recovery.

**International Commission for the Conservation of Atlantic Tuna**
Principe de Vergara 17-7
E-28001 Madrid, Spain
431 03 29

**International Council for Bird Preservation**
32 Cambridge Road
Girton, Cambridge CB3 OPJ
 England
0223 277318
*Russell W. Peterson, President*

Reports on status of bird species worldwide and initiates conservation projects.

**International Council for Game and Wildlife Conservation**
15 rue de Teheran
F-75008 Paris, France
45 63 51 33

Promotes hunting in harmony with nature conservancy.

**International Crane Foundation**
E-11376, Shady Lane Road
Baraboo, WI 53913-9778
(608) 356-9462
*Claire Mirande, Curator of Birds*

Works on habitat preservation, propagation, and re-stocking of cranes.

**International Ecology Society**
1471 Barclay Street
St. Paul, MN 55106-1405
(612) 774-4971
*R. J. F. Kramer, President*

International organization for the protection of the environment and a better understanding of all life forms.

**International Fund for Animal Welfare**
P.O. Box 193
Yarmouth Port, MA 02675
(508) 362-4944
*Brian D. Davies, Founder*

Dedicated to the protection of wild and domestic animals.

## International Game Fish Association
3000 E. Las Olas Boulevard
Ft. Lauderdale, FL 33316-1616
(305) 467-0161
*Elwood K. Harry, President*

Desires ethical international angling regulations.

## International North Pacific Fisheries Commission
6640 NW Marine Drive
Vancouver, British Columbia
1X2 V6T Canada
(604) 228-1128
*B. E. Skud, Executive Director*

Canada, Japan, and the U.S. joining to conserve fisheries in the North Pacific Ocean.

## International Osprey Foundation
P.O. Box 250
Sanibel Island, FL 33957
(813) 472-5218
*David Loveland, President*

Goal is to restore osprey to a stable population.

## International Pacific Halibut Commission
P.O. Box 95009
Seattle, WA 98145-2009
(206) 634-1838
*Donald A. McCaughran, Director*

Joint commission, U.S. and Canada, to research and preserve Pacific halibut.

## International Primate Protection League
P.O. Box 766
Summerville, SC 29484
(803) 871-2280
*Shirley McGreal, Chairperson*

Protects all species of primates.

## International Professional Hunters' Association
P.O. Box 17444
San Antonio, TX 78217
(512) 824-7506
*Donald Lindsay, President*

Encourages fair chase and sportsman-like conduct in sport hunting.

## International Snow Leopard Trust
16463 S.E. 35th Street
Bellevue, WA 98008
(206) 632-2421

Protects snow leopards in the wild and in captivity.

## International Society for the Protection of Mustangs and Burros
11790 Deodar Way
Reno, NV 89506
(702) 972-1989
*Helen A. Reilly, President*

Protects and preserves the animals and their habitats.

## International Waterfowl and Wetlands Research Bureau
Simbridge
Gloucester GL2 7BX England
389 333
389 827 FAX

## International Whaling Commission
The Red House, 135 Station Road
Histon, Cambridge CB4 4NP
England
0223 233971
*S. Irberger, Chairman (Sweden)*

Interested in the conservation of whale stocks and the development and regulation of the whaling industry.

## International Wild Waterfowl Association
R.F.D. #1, James Farm
Durham, NH 03824
(603) 659-5442
*Walter B. Sturgeon, Jr., President*
Protects all species of waterfowl.

## International Wildlife Coalition
634 North Falmouth Highway,
P.O. Box 388
North Falmouth, MA 02556-0388
(508) 654-9980
*Daniel J. Morast, Chairman*

Sponsors an "Adopt a Whale" program and publishes both an adult version and children's version of their newsletter.

## International Wildlife Rehabilitation Council
1171 Kellogg Street
Suisun, CA 94585
(707) 428-IWRC
*Kris Thorne, President*

Made up of organizations that handle sick and injured wild animals.

## Irish Wildlife Federation
Ferry House, 4th Floor, Lower
Mount Street
Dublin 2, Ireland
60 83 46

## Izaak Walton League of America
1401 Wilson Boulevard, Level B
Arlington, VA 22209
(703) 528-1818
*Donald L. Ferris, Chairman,
Executive Board*

National organization of sportfishermen with general conservation interests.

South Dakota Division
711 Franklin Street
Rapid City, SD 57701
(605) 342-3256

Virginia Division
205 Maple Street
Fredericksburg, VA 22405
(703) 373-4486
*Samuel Mason, President*

West Virginia Division
200 Lakeside Drive, G-6
Morgantown, WV 26505
(304) 296-2903
*Rosemary Wilson, President*
*Marie Cyphert, Secretary*

Wisconsin Division
1003 4th Street
Stevens Point, WI 54481
(414) 242-0813
*Gerald Ernst, President*

Wyoming Division
1072 Empinado
Laramie, WY 82070
(307) 742-2785
*Raymond G. Jacquot, President*

## Jack Miner Migratory Bird Foundation
P.O. Box 39
Kingsville, Ontario N9Y 2E8
Canada
(519) 733-4034
*Jasper W. Miner, President*

A public foundation founded in 1904 to protect migratory birds in the U.S. and Canada.

## Kangaroo Protection Foundation
1807 H Street, NW
Washington, DC 20006
(202) 347-0822
*Marian Newman, Program Director*

Promotes the conservation and humane treatment of kangaroos, wallabies, and other land mammals.

## Kansas Department of Wildlife and Parks
900 Jackson Street, Suite 502
Topeka, KS 66612-1220
(913) 296-2281
*Robert L. Meinen, Secretary*

## Kentucky Department of Fish and Wildlife Resources

#1 Game Farm Road
Frankfort, KY 40601
(502) 564-3400
*Don R. McCormick, Commissioner*

## Laboratory of Ornithology

Cornell University
159 Sapsucker Woods Road
Ithaca, NY 14850
(607) 254-2473
*Charles Walcott, Executive Director*

Center for the study of birds.

## Last Chance Forever

506 Avenue A
San Antonio, TX 78215
(512) 655-6049
*John Karger, Executive Director*

Shelters and rehabilitates birds of prey.

## Louisiana Department of Wildlife and Fisheries

P.O. Box 98000
Baton Rouge, LA 70898
(504) 765-2800
*Virginia Van Sickle, Secretary*

## Maine Atlantic Sea Run Salmon Commission

P.O. Box 1298, Hedin Hall
Bangor, ME 04401
(207) 941-4449
*William J. Vail, Chairman*

## Maine Department of Inland Fisheries and Wildlife

284 State Street, Station #41
Augusta, ME 04333
(207) 289-2766
*William J. Vail, Commissioner*

## Manitoba Wildlife Federation

1770 Notre Dame Avenue
Winnipeg, Manitoba R3E 3K2
Canada
(204) 633-5967
*Robert Barton, President*

## Manomet Bird Observatory

P.O. Box 936
Manomet, MA 02345
(508) 224-6521
*William S. Brewster, Chairman*

Research and education institute specializing in shorebird migration and wetlands.

## Marine Mammal Commission

1625 I Street, NW
Washington, DC 20006
(202) 653-6237
*John R. Twiss, Jr., Executive Director*

Reviews the status of marine mammal populations.

## Mariposa Monarca

Avenida Constituyentes 345-806,
Colonia Danile Garza
11830, Mexico City, D.F., Mexico
534-1023

Protects the Monarch butterfly

## Massachusetts Department of Fisheries, Wildlife and Environmental Law Enforcement

100 Cambridge Street
Boston, MA 02202
(617) 727-1614
*Walter E. Bickford, Commissioner*

## Max McGraw Wildlife Foundation

P.O. Box 9
Dundee, IL 60118
(312) 741-8000
*Stanley Koenig, Executive Director*

Wildlife and fishery research center.

**Memorial Wildlife Federation**
P.O. Box 240
Holbrook, NY 11741
(516) 567-0031
*Stanley Bethiel, President*

A scientific organization dedicated to the protection of wildlife.

**Michigan, University of**
Resource Ecology and
    Management
Ann Arbor, MI 48109-1115
(313) 764-2550
*Gary W. Fowler, Coordinator*

Publishes *Endangered Species UP-DATE.*

**Migratory Bird Conservation Commission**
Interior Building
Washington, DC 20240
(202) 653-7653
*Manuel Lujan, Jr., Secretary of the Interior, Chairman*

Purchases migratory bird habitats for the National Wildlife Refuge System.

**Minnesota Herpetological Society**
James Ford Bell Museum of
    Natural History
10 Church Street, SE, University of
    Minnesota
Minneapolis, MN 55455-0104
(612) 626-2030
*John Moriarity, President*

Conserves and preserves reptiles and amphibians.

**Mississippi Department of Wildlife, Fisheries and Parks**
Southport Mall, P.O. Box 451
Jackson, MS 39205
(601) 961-5300
*Vernon Bevill, Executive Director*

**Mlilwane Game Sanctuary**
P.O. Box 33
Mbabane, Swaziland
61037
*T. E. Reilly, Manager*

**Monitor**
1506 19th Street, NW
Washington, DC 20036
(202) 234-6576
*Craig Van Note, Executive Vice-President*

Information clearinghouse on marine animals.

**Montana Department of Fish, Wildlife and Parks**
1420 East Sixth
Helena, MT 59620
(406) 444-2535
*K. L. Cool, Director*

**Mountain Gorilla Project**
1717 Massachusetts Avenue, NW, Suite 602
Washington, DC 20036
(202) 265-8394
*Diana McMeekin, Vice-President*

A project of the African Wildlife Foundation. Works with Rwandan authorities to support conservation.

**Muskies, Inc.**
2301 7th Street
North Fargo, ND 58102
(701) 232-9544
*Steve Statland, President*

Introduces muskellunge into suitable waters and fights water pollution.

**National American Falconers Association**
P.O. Box 63
Winifred, MT 59489
(406) 462-5487
*Ralph R. Rogers, President*

## National Coalition for Marine Conservation

P.O. Box 23298
Savannah, GA 31403
(912) 234-8062
*Christopher M. Weld, President*

Conserves ocean fish and their environment.

## National Field Archery Association

31407 Outer I-10
Redlands, CA 92373
(714) 794-2133
*John Slack, President*

Archers and bowhunters working toward conservation of game in its natural habitat.

## National Fish and Wildlife Foundation

Main Interior Building
18th and C Streets, NW, Room 2556
Washington, DC 20240
(202) 343-1040
*Charles H. Collins, Executive Director*

Established by the U.S. Congress to encourage the private sector to contribute to wildlife programs.

## National Flyway Council

Wyoming Game and Fish Department
Cheyenne, WY 82002
(307) 777-7604
*Dale Strickland, Chairman*

Protects the migration routes of various birds.

Atlantic Flyway
50 Wolf Road
Albany, NY 12233
(518) 457-5690
*Kenneth Wich, Director*

Mississippi Flyway
500 Lafayette Road
St. Paul, MN 55155
*Roger Holmes, Chairman*

Pacific Flyway
1420 East 6th Avenue
Helena, MT 59620
(406) 444-3186
*Ron Marcoux, Associate Director*

## National Foundation to Protect America's Eagles

Save the Eagle Project
P.O. Box 120206
Nashville, TN 37212
(615) 847-4171
*Al Louis Cecere, President*

Preserves and protects the endangered bald eagle.

## National Hunters Association

P.O. Box 820
Knightdale, NC 27545
(919) 365-9289
*D. V. Smith, President*

Protects the rights of hunters.

## National Institute for Urban Wildlife

10921 Trotting Ridge Way
Columbia, MD 21044-2831
(301) 596-3311
*Gomer E. Jones, President*

Studies man's relationship to wildlife in an urban environment.

## National Military Fish and Wildlife Association

Code 241B, Atlantic Division
NAVFA-CENGCOM,
Norfolk, VA 23511-6287
(804) 445-2369
*Slader Buck, President*

Organization of natural resources personnel within the Department of Defense.

**National Prairie Grouse Technical Council**
Michigan Department of Natural Resources
Rt. 2, Box 2555
Manistique, MI 49854
(906) 341-6917
*Greg Stoll, Chairman*

Preserves and researches the prairie chicken and the sharp-tailed grouse.

**National Trappers Association**
216 North Center Street,
P.O. Box 3667
Bloomington, IL 61702
(309) 829-2422
*Scott Hartmann, President*

Trappers organization promoting environmental·programs along with an annual fur bearer harvest.

**National Wild Turkey Federation**
Wild Turkey Building,
P.O. Box 530
Edgefield, SC 29824-0530
(803) 637-3106
*Rob Keck, Executive Vice-President*

Conserves the American Wild Turkey.

**National Wildlife Rehabilitators Association**
R.R. 1, Box 125E
Brighton, IL 62012
(618) 372-3083
*Elaine M. Thrune, President*

Cares for and releases distressed wild animals.

**Nebraska Game and Parks Commission**
2200 North 33rd Street, P.O. Box 30370
Lincoln, NE 68503
(402) 464-0641
*Rex Amack, Director*

**Nevada Department of Wildlife**
P.O. Box 10678
Reno, NV 89520
(702) 789-0500
*William A. Molini, Director*

**Nevada State Commission for the Preservation of Wild Horses**
Stuart Facility
Capitol Complex
Carson City, NV 89710
(702) 885-5589

**New Brunswick Wildlife Federation**
190 Cameron Street
Moncton, New Brunswick
5Z2 E1C Canada
(506) 857-2056
*James Marriner, President and Executive Director*

**New Hampshire Fish and Game Department**
2 Hazen Drive
Concord, NH 03301
(603) 271-3421
*Donald A. Normandeau, Executive Director*

**New Jersey Audubon Society**
P.O. Box 125, 790 Ewing Avenue
Franklin Lakes, NJ 07417
(201) 891-1211
*Fred L. Ditmars, President*

Organized in 1897 to foster environmental awareness. Not affiliated with the National Audubon Society.

**New York State Fish and Wildlife Management Board**
Albany, NY 12233
(914) 761-2653
*Roger H. Cole, Chairman*

Membership composed of government leaders, sportsmen, and landowners.

## New York Turtle and Tortoise Society
163 Amsterdam Ave.
Suite 365
New York, NY 10023
(212) 459-4803

Veterinarians, zookeepers, biologists, and pet owners dedicated to the appreciation of turtles and tortoises—their conservation, preservation, and propagation.

## New York Zoological Society
The Zoological Park
Bronx, NY 10460
(212) 220-5100
*Howard Phipps, Jr., President*

## North American Blue Bird Society
P.O. Box 6295
Silver Spring, MD 20906
(301) 384-2798
*Sadie Dorber, President*

## North American Loon Fund
R.R. 4, Box 240C
Meredith, NH 03253
(603) 279-6163
*Scott Sutcliffe, Chairman*

## North American Native Fishes Association
123 West Mt. Airy Avenue
Philadelphia, PA 19119
(215) 247-0384
*Bruce Gebhardt, President*

Works to increase the appreciation of native species.

## North American Wildlife Foundation
102 Wilmot Road, #410
Deerfield, IL 60015
(312) 940-7776
*P. A. W. Green, President*

Owns Delta Waterfowl Research Station and works to benefit wildlife and other natural resources.

## North American Wildlife Park Foundation
Wolf Park
Battle Ground, IN 47920
(317) 567-2265
*Erich Klinghammer, President*

Operates Wolf Park and conducts ongoing research into animal behavior, with a concentration on wolves.

## North American Wolf Society
P.O. Box 82950
Fairbanks, AK 99708
(907) 474-6117
*Anne K. Ruggles, President*

Encourages a rational approach to the conservation of wolves and other wild canids.

## North Atlantic Salmon Conservation Organization
11 Rutland Square
Edinburgh EH1 2AS United Kingdom
228 2551

## North Carolina Wildlife Resources Commission
Archdale Building, 512 North Salisbury Street
Raleigh, NC 27611
(919) 733-3391
*Charles R. Fullwood, Executive Director*

## North Dakota State Game and Fish Department
100 North Bismarck Expressway
Bismarck, ND 58501
(701) 221-6300
*Lloyd A. Jones, Commissioner*

## North Pacific Fur Seal Commission

c/o Marine Fisheries Service
1335 East-West Highway
Silver Spring, MD 20910
(301) 427-2239
*James W. Brennan, Assistant Administrator*

Works to maintain the maximum sustainable productivity of fur seal resources.

## North-East Atlantic Fisheries Commission

Room 425, Nobel House, 17 Smith Square
London SW1P 3HX England
01-238 5919
*W. Ranke, President*

Goals are the conservation and optimum utilization of the fishery resources of the Northeast Atlantic.

## Northwest Atlantic Fisheries Organization

P.O. Box 638
Dartmouth, Nova Scotia B2Y 3Y9
Canada
(902) 469-9105
*Joaquim C. Esteves Cardoso, Executive Secretary*

## Oklahoma Department of Wildlife Conservation

1801 North Lincoln, P.O. Box 53465
Oklahoma City, OK 73152
(405) 521-3851
*Steven Alan Lewis, Director*

## Oregon Department of Fish and Wildlife

2501 SW First Avenue
Portland, OR 97201
(503) 229-5551
*Randy Fisher, Director*

## Organization of Wildlife Planners

Box 7921
Madison, WI 53707
(307) 777-7461
*Walt Gasson, President*

Nationwide coalition of professionals involved in wildlife planning.

## Pacific Seabird Group

4990 Shoreline Highway
Stinson Beach, CA 94970
(916) 752-1300
*Michael Fry, Chair*

Studies and conserves the over 275 species of Pacific seabirds.

## Pacific States Marine Fisheries Commission

2501 SW 1st Avenue, Suite 200
Portland, OR 97201-4752
(503) 326-7025
(503) 326-7033 FAX
*Guy Thornburgh, Executive Director*

Commission serves the states of Alaska, California, Idaho, Oregon, and Washington to promote conservation and management of marine fisheries.

## Pacific Whale Foundation

101 N. Kihei Road
Kihei, HI 96753
(808) 879-8811
*Gregory Kaufman, President*

Coalition of scientists, conservationists, and volunteers united to prevent the extinction of marine mammals.

## Pennsylvania Fish Commission

P.O. Box 1673
Harrisburg, PA 17105
(717) 657-4518
*David Coe, President*

## Pennsylvania Game Commission
2001 Elmerton Avenue
Harrisburg, PA 17110-9797
(717) 787-4250
*Peter S. Duncan III, Executive Director*

## Peregrine Fund, The
5666 West Flying Hawk Lane
Boise, ID 83709
(208) 362-3716
*Tom J. Cade, Founder*

Founded in 1970 to protect the falcon.

## Pheasants Forever
P.O. Box 75473
St. Paul, MN 55175
(612) 481-7142
*Garu C. Molitor, President*

Group dedicated to the preservation and propagation of pheasants.

## Porpoise Rescue Foundation
405 West Washington Street, Box 162
San Diego, CA 92103
(619) 295-5682
*Harold Cary, President*

Established by the U.S. Tuna Foundation to work with tuna boat skippers to develop effective porpoise rescue methods.

## Professional Bowhunters Society
P.O. Box 5275
Charlotte, NC 28225
(704) 536-6009
*Fred Richter, President*

Seeks to upgrade and preserve the sport of bowhunting.

## Protect Our Pelican Society
P.O. Box 2648
Sarasota, FL 34230
(813) 955-2266
*Dale Shields, President and Founder*

Supports environmental and wildlife preservation and operates Pelican Man's Bird Sanctuary.

## Purple Martin Conservation Association
Institute for Research and Community Services
Edinboro University of Pennsylvania
Edinboro, PA 16444
(814) 734-4420
*James R. Hill III, Director*

Conducts research and conservation of wild birds.

## Quail Unlimited
P.O. Box 10041
Augusta, GA 30903
(803) 637-5731
*M. McNiell Holloway, President*

Works for quail and upland bird conservation through habitat management.

## Ranger Rick's Nature Club
8925 Leesburg Pike
Vienna, VA 22180
(703) 790-4000
*Gerald Bishop, Editor*

Division of the National Wildlife Federation whose purpose is to teach children to know and respect all living things. Membership of more than 700,000.

## Raptor Education Foundation
21901 East Hampden Avenue
Aurora, CO 80013
(303) 680-8500
*Peter Reshetniak, President*

Informs the public on the importance of preserving birds of prey.

## RARE
19th and the Parkway
Philadelphia, PA 19103
(215) 299-1182
*George S. Glenn, Jr., Executive Director*

Rare Animal Relief Effort focusing on tropical birds.

## Remington Farms
R.D. #2, Box 660
Chestertown, MD 21620
(301) 778-1565
*E. Hugh Galbreath, Manager,
Wildlife Management*

Operated by Remington Arms Co., a habitat designed to "farm" wildlife.

## Rocky Mountain Bighorn Society
Colorado Order
P.O. Box 1086
Denver, CO 80201
(303) 278-2502
*Tom Brown, President*

Supports the sound management of sheep and their habitats.

## Rocky Mountain Elk Foundation
P.O. Box 8249
Missoula, MT 59802
(406) 721-0010
*Wallace Pate, Chairman of the
Board*

Promotes the preservation of elk and their habitats.

## Roo Rat Society
Whitman College
Walla Walla, WA 99362
(509) 527-5229
*James S. Todd, Scribe*

Conservation organization, derives its name from the kangaroo rat.

## Ruffed Grouse Society
1400 Lee Drive
Coraopolis, PA 15108
(412) 262-4044
*Samuel R. Pursglove, Jr., Executive
Director*

Dedicated to improving the environment of the ruffed grouse, woodcock, and other forest wildlife.

## Safari Club International
4800 West Gates Pass Road
Tucson, AZ 85745
(602) 620-1220
*Warren K. Parker, President*

Sportsmen united to encourage conservation of wildlife.

## Saga Furs of Scandinavia
156 5th Avenue, #308
New York, NY 10010
(212) 463-7670
*Lili Glassman, Spokesperson*

Fur trade association.

## San Diego Zoo
P.O. Box 551
San Diego, CA 92112
(619) 234-3153
*Doug Myers, Executive Director*

The largest and most popular zoo in America.

## Saskatchewan Natural History Society
Box 414
Raymore, Saskatchewan SOA 3JO
Canada

Involved in sanctuaries and educational publications.

## Saskatchewan Wildlife Federation
Box 788
Moose Jaw, Saskatchewan S6H 4P5 Canada
(306) 692-7772
*Ed Begin, Executive Director*

## Save the Manatee Club
500 North Maitland Avenue
Maitland, FL 32751
(407) 539-0990
*Jimmy Buffett, Chairman,
Co-founder*

Promotes public awareness of the endangered West Indian manatee.

## Save the Whales
P.O. Box 3650
Washington, DC 20007
(202) 337-2332
*Christine Stevens, President*

Informs the public about the dilemma of the great whales, works to save them from extinction, and opposes commercial whaling.

## Sea Shepherd Conservation Society
P.O. Box 7000-S
Redondo Beach, CA 90277
(213) 373-6979
*Paul Watson, Founder*

Protects and conserves marine mammals.

## Smallmouth, Inc.
260 Crest Road
Edgefield, SC 29824
(803) 637-5722
*Tom Rodgers, President*

Researches smallmouth bass and represents the interests of fishermen.

## Smithsonian Institution
National Zoological Park
3000 Connecticut Avenue, NW
Washington, DC 20008
(202) 357-2700
*Michael Hill Robinson, Director*

## Society for Animal Protective Legislation
P.O. Box 3719, Georgetown Station
Washington, DC 20007
(202) 337-2334
*Medeleine Bemelmans, President*

Instrumental in obtaining enactment of 14 federal laws for the protection of animals.

## Society for Marine Mammalogy
c/o Sea World Research Institute
1700 South Shores Road
San Diego, CA 92109
(619) 226-3877
*Randall W. Davis, Secretary*

Conserves and researches marine mammals.

## Society for the Preservation of Bighorn Sheep
3113 Mesaloa Lane
Pasadena, CA 91107
(818) 797-1287
*George Kerr, President*

Studies, researches, and preserves the bighorn sheep.

## Society for the Preservation of Birds of Prey
P.O. Box 66020
Los Angeles, CA 90066
(213) 397-8216
*J. Richard Hilton, President*

Seeks to advocate the strictest possible protection of birds of prey.

## Society of Tympanuchus Cupido Pinnatus Ltd.
930 Elm Grove Road
Elm Grove, WI 53122
(414) 782-6333
*Bernard J. Westfahl, President*

Preserves the prairie chicken.

## South Carolina Wildlife and Marine Resources Department
Rembert C. Dennis Building,
P.O. Box 167
Columbia, SC 29202
(803) 734-3888
*James A. Timmerman, Jr.,
Executive Director*

**South Dakota Game, Fish and Parks Department**
445 East Capitol
Pierre, SD 57501-3185
(605) 773-3718
*Richard Beringson, Secretary*

**Sport Fishery Research Program**
1010 Massachusetts Avenue, NW
Washington, DC 20001
(202) 898-0771
*D. F. Myers, President*

Helps finance the graduate level training of fishery scientists and to support research in the sport fishery resources field.

**Sportsmen Conservationists of Texas**
311 Vaughn Building, 807 Brazos Street
Austin, TX 78701
(512) 472-2267
*Tom Martine, President*

Affiliated with the National Wildlife Federation.

**Tennessee Conservation League**
11 Music Circle South, Suite 5
Nashville, TN 37203
(615) 254-7364
*Larry Richardson, President*

Promotes conservation and wildlife appreciation.

**Tennessee Wildlife Resources Agency**
P.O. Box 40747, Ellington Agricultural Center
Nashville, TN 37204
(615) 781-6500
*Gary Myers, Executive Director*

Has exclusive jurisdiction over wildlife, boating, hunting and fishing in Tennessee.

**Texas Armadillo Association**
P.O. Box 311074
New Braunfels, TX 78131
(512) 629-4980
*Jim Schmidt, Executive Officer*

Promotes the well-being of armadillos by sponsoring a worldwide traveling show and media appearances of armadillos.

**Texas Parks and Wildlife Department**
4200 Smith School Road
Austin, TX 78744
(512) 389-4800
*Charles D. Travis, Executive Director*

**Texas Wildlife Association**
1635 NE Loop 410, Suite 108
San Antonio, TX 78209
(512) 826-2904
*Richard M. Butler, President*

Formed to promote the rights of sportsmen, wildlife managers and land operators.

**TRAFFIC (USA)**
1250 24th Street, NW
Washington, DC 20037
(202) 293-4800
*Ginette Hemley, Director*

Monitors world trade in wild animals and fauna.

**Trout Unlimited**
501 Church Street, NE, Suite 103
Vienna, VA 22180
(703) 281-1100
*Robert L. Herbst, Executive Director*

Founded to preserve and enhance the coldwater habitat of trout, salmon, and steelhead bass.

Rhode Island Chapter
(Narragansett)
492 Cumberland Avenue
North Attleboro, MA 02760
(617) 695-3288

Tennessee Council
106 Scotch Street
Hendersonville, TN 37075
(615) 824-5140
*Mark Lamberth, Chairman and
National Director*

Wisconsin Council
N3079 Tomlinson Road
Foynette, WI 53955
(608) 241-3311
*Tom Flesch, National Director*

Wyoming
P.O. Box 2185
Jackson, WY 83001
(307) 733-5261
*Edward Ingold, Chairman and
Director*

**Trumpeter Swan Society**
3800 County Road 24
Maple Plane, MN 55359
(612) 476-4663
*David K. Weaver, Executive
Secretary-Treasurer*

Advocates restoration of trumpeter swans in their original range.

**United Nations Secretariat on
the Conservation of
Migratory Species of Wild
Animals**
Wissenschaftszentrum, Ahrstrsse 45
D-5300 Bonn 2, Germany
302 152

**Washington Department of
Fisheries**
115 General Administration
Building
Olympia, WA 98504
(206) 753-6600
*Joseph R. Blum, Director
Judith Merchant, Deputy Director*

Created in 1890.

**Washington Department of
Wildlife**
600 Capitol Way
North Olympia, WA 98501-1091
(206) 753-5700
*Curt Smith, Director*

**Waterfowl Habitat Alliance of
Texas**
1973 West Gray, Suite 6
Houston, TX 77019
(713) 522-5025
*Richard Tinsley, Chairman*

Also known as WHAT Ducks. Coordinates statewide duck box program.

**Welder Wildlife Foundation**
P.O. Box 1400
Sinton, TX 78387
(512) 364-2643
*James G. Teer, Director*

Does conservation and research in wildlife ecology.

**Western Association of Fish
and Wildlife Agencies**
P.O. Box 944209
Sacramento, CA 94244-2090
(916) 323-7319
*Jerry M. Conley, President*

Organization of 13 states and three Canadian provinces.

**Western Bird Banding
Association**
3975 North Pontatoc
Tucson, AZ 85718
(602) 299-1287
*Robert C. Tweit, President*

Promotes research and conservation through bird banding.

**Wetlands for Wildlife**
P.O. Box 344
West Bend, WI 53095
(414) 628-1060
*Lambert Neuberg, President*

Preserves and acquires wetlands and wildlife habitats.

## Whale Center
3929 Piedmont Avenue
Oakland, CA 94611
(415) 654-6621
*Mark J. Palmer, Executive Director*

Individuals interested in whales and whale conservation.

## Whitetails Unlimited
P.O. Box 422
Sturgeon Bay, WI 54235
(414) 743-6777
*Jeffrey B. Schinkten, President*

Preserves the whitetail deer.

Illinois
R.R. 2, Box 331
Athens, IL 62613
*Larry Yoakum, District Representative*

New York
74 Mohawk Street
Fort Plain, NY 13339
*Gerald Hudson, District Representative*

Wisconsin
Marshfield, WI 54449
*Marilyn Laidlaw, District Representative*

## Whooping Crane Conservation Association
3000 Meadowlark Drive
Sierra Vista, AZ 85635
(602) 458-0971
*Lawrence S. Smith, President*

Naturalists, ornithologists, and aviculturists working to prevent the extinction of the whooping crane.

## Wild Canid Survival and Research Center/Wolf Sanctuary
P.O. Box 760
Eureka, MO 63025
(314) 938-5900
*Gary Schoenberger, President*

Maintains endangered wolves for possible re-establishment efforts.

## Wild Horse Organized Assistance (WHOA)
P.O. Box 555
Reno, NV 89504
(702) 323-5908
*Dawn Y. Lappin, Executive Director*

A foundation for the welfare and perpetuation of the wild free-roaming horses and burros.

## Wildfowl Foundation
2111 Jefferson Davis Highway, 605-S
Arlington, VA 22202
(703) 979-2626
*C. R. Gutermuth, President*

Conserves ducks, geese, and swans.

## Wildfowl Trust of North America
P.O. Box 519
Grasonville, MD 21638
(301) 827-6694
*Steven F. Capranica, President*

Fosters public stewardship of wetlands and waterfowl.

## Wildlife Clubs of Kenya
P.O. Box 40658
Nairobi, Kenya
74265

## Wildlife Conservation International (WCI)
New York Zoological Society
185th Street and South Boulevard, Building A
Bronx, NY 10460
(212) 220-5155
*David Western, Director*

Established in 1895, the oldest international wildlife organization in the U.S.

## Wildlife Disease Association

P.O. Box 886
Ames, IA 50010
(515) 233-1931
*Ed Addison, President*

Organization of scientists interested in advancing knowledge of disease and how environmental factors affect wildlife.

## Wildlife Education Ltd.

P.O. Box 85271, Suite 6
San Diego, CA 92138
(800) 334-8152

Publications for children.

## Wildlife Education Program for a Living Future

Bittner's Resort
H.C. 2, Box 225A
Bovey, MN 55709
(218) 245-3049
*Karlyn Atkinson Berg, Director*

Promotes understanding and conservation of wolves.

## Wildlife Games

P.O. Box 247
Ivy, VA 22945
(804) 972-7016

Created and distributes "Whale Fact," a game where children become a migrating whale.

## Wildlife Information Center

629 Green Street
Allentown, PA 18102
(215) 434-1637
*Donald S. Heintzelman, President*

Promotes non-killing wildlife activities such as photography and observation.

## Wildlife Legislative Fund of America and Wildlife Conservation Fund of America

50 West Broad Street
Columbus, OH 43215
(614) 221-2684
*Daniel M. Galbreath, Chairman*

Companion organizations established to protect American sportsmen's interests.

## Wildlife Management Institute

1101 14th Street, NW, Suite 725
Washington, DC 20005
(202) 371-1808
*Laurence R. Jahn, President*

Promotes improved professional management of natural resources.

## Wildlife Preservation Trust International

34th Street and Girard Avenue
Philadelphia, PA 19104
(215) 222-3636
*Gerald M. Durrell, Founder and Honorary Chairman*

Supports captive propagation.

## Wildlife Society

5410 Grosvenor Lane
Bethesda, MD 20814
(301) 897-9770
*Richard J. Mackie, President*

International organization of professionals and students engaged in wildlife research, management, education, and administration. Founded 1937.

Central Mountains and Plains Section
c/o Wildlife Research Center,
317 West Prospect
Fort Collins, CO 80526
(303) 484-2836
*Clait E. Braun, Representative*

North Central Section
c/o Wisconsin Cooperative
  Wildlife Research Unit
211 Russell Labs, University of
  Wisconsin
Madison, WI 53706
(608) 263-6882
*Donald H. Rusch, Representative*

Northeast Section
P.O. Box 225
Leverett, MA 01054
(413) 549-0520
*William M. Healy,*
  *Representative*

Northwest Section
c/o Department of Fisheries and
  Wildlife
Oregon State University, 104
  Nash Hall
Corvallis, OR 97331-3803
(503) 754-4531
*Robert G. Anthony,*
  *Representative*

Southeast Section
c/o Natural Resources and
  Rural Development Unit,
Room 3871, South Building
Washington, DC 20250
(202) 447-5468
*James E. Miller, Representative*

Southwest Section
Department of Wildlife and
  Fisheries Sciences
Texas A&M University
College Station, TX 77843
(409) 845-5777
*Nova J. Silvy, Representative*

Western Section
Forestry Sciences Lab, 2081 East
  Sierra Avenue
Fresno, CA 93710
(209) 487-5589
*John G. Kie, Representative*

**Wildlife Way Station**
14831 Little Tujunga Canyon
Angeles National Forest,
  CA 91342
(818) 899-5201
*Martine Colette, President*

Station cares for abandoned, sick, and homeless wild animals. Founded by actress Tippi Hedren.

**William Holden Wildlife Association**
P.O. Box 67981
Los Angeles, CA 90067
(213) 274-3169
*Stefanie Powers, Founder*

Supports worldwide conservation of wildlife, has been extremely active in Africa.

**Wilson Ornithological Society**
c/o President—Department of
  Ornithology
Royal Ontario Museum
Toronto, Ontario M5S 2C6 Canada
(416) 586-5522
*Jon C. Barlow, President*

Organized in 1888 to advance the science of ornithology.

**World Pheasant Association**
P.O. Box 5, Lower Basildon
Reading RG8 9PFT England
5140

Group dedicated to the preservation and propagation of pheasants.

**World Pheasant Association of U.S.A.**
c/o Brookfied Zoo
3300 Golf Road
Brookfield, IL 60513
(312) 485-0263
*Edward C. Schmitt, President*

## World Underwater Foundation
34 rue de Colisee
F-75008 Paris, France
42 25 85

Organization of competitors in underwater spearfishing.

## World Wide Fund for Nature
Avenue du Mont-Blanc
CH-1196 Gland, Switzerland
64 71 81
64 46 15
*Prince Philip, Duke of Edinburgh, President*

Also known as the World Wildlife Fund. The largest private international nature conservation organization, with more than three million supporters on all continents.

Denmark
Osterbrogade 94
DK-2100 Copehagen 0, Denmark
38 20 20

India
c/o Godrej & Boyce Ltd., Lalbaug, Parel
Bombay 400 012, India
413 2927

Italy (Fondo Mondiale per la Natura—Italia)
Via Salaria 290
I-00199 Rome, Italy
85 24 92

Japan
Nihonseimei Akabanebashi Building 7F, 3-1-14 Shiba
Minatoku, Tokyo 105, Japan
769 1711

Madagascar (Fonds Mondial pour la Nature)
B.P. 4373
Antananarivo, Madagascar
25541

Malaysia
8th Floor, Wisma Damansara, Jalan Semantan, P.O. Box 10769
50724 Kuala Lumpur, Malaysia
255 44 95

Netherlands
Postbus 7
NL-3700 AA Zeist, The Netherlands
22 164

New Zealand
35 Taranaki Street, 2nd Floor, P.O. Box 6237
Wellington, New Zealand

Norway
Hegdebaugsveien 22
N-0167 Oslo 1, Norway
69 61 97

Pakistan
P.O. Box 1312
Lahore, Pakistan
85 11 74

South Africa
P.O. Box 456
Stellenbosch 7600, South Africa
72892

Sweden
Ulriksdals Slott
S-171 71 Solna, Sweden
85 01 20

Switzerland (Fonds Mondial pour la Nature)
Forrlibuckstrasse 66, Postfach 749
CH-8037 Zurich, Switzerland
44 20 44

## World Wildlife Fund
1250 24th Street, NW
Washington, DC 20037
(202) 293-4800
*Kathryn Fuller, President*
*Russell Train, Chairman*

Seeks to protect the biological resources upon which human well-being depends.

Australia
St. Martins Tower, Level 17
31 Market Street, GPO Box 528
Sydney, New South Wales 2001
Australia
261 5572

Canada
20 St. Clair Avenue, East, Suite
201
Toronto, Ontario M4T 1N5
Canada
(416) 923-8173

**Wyoming Game and Fish Department**
5400 Bishop Road
Cheyenne, WY 82002
(307) 777-7735
*Francis E. Petera, Director*

**Xerces Society, The**
10 SW Ash Street
Portland, OR 97204
(503) 222-2788
*Edward O. Wilson, President*

Group of scientists working in conservation-related fields, particularly butterfly protection. Named after an extinct San Francisco butterfly, the Xerces Blue.

# APPENDIX

## ECO-FACTS

- Lighting in the U.S. accounts for 25% of our total electrical use.

- Each day every one of us throws away 3½ pounds of trash. By the end of a year that's almost 1,300 pounds each.

- Each morning 50,000 trash trucks roll out to haul away your trash.

- Approximately 11% of waste is recycled now.

- An average family of 3 produces about 5 pounds per week, 20 pounds per month, or 250 pounds per year of used newspaper.

- Each year the United States uses 85.5 million tons of paper—we recycle 22% or 19 million tons.

- Nearly 70% of the earth's surface is covered by water.

- The United States uses about 400 billion gallons of water each day, or about 1,650 gallons for each person each day.

- Already costing us more than $10 billion a year, trash disposal is only getting more expensive.

- Most packaging goes straight from the store shelf to your wastebasket, accounting for a full third of the waste we produce.

- 40% of the energy you use in your home is for heat.

- Almost half of all energy used in our homes leaks out windows, attics, or cracks.

- Our refrigerators consume 7% of the nation's electricity.

- As much as 90% of the energy used by washing machines is for heating the water.

- Washers take up to 59 gallons of water to use.

- A drip from a leaky faucet can waste over 50 gallons of water a day.

- Washing the car with a running hose uses up to 150 gallons of water; a sponge and bucket takes 15 gallons.

- Americans produce 160 million tons of garbage every year—enough to fill the New Orleans Superdome twice a day, every day.

- If Americans recycled all their Sunday papers, 500,000 trees would be saved each week.

• Plastic waste kills up to a million sea birds, 100,000 sea mammals, and innumerable fish each year.

# WATER FACTS

• How much water is used . . . ?

1. In the average residence during a year? 107,000 gallons
2. By an average person daily? 168 gallons
3. To flush a toilet? 5–7 gallons
4. To take a shower? 25–50 gallons
5. To brush your teeth (water running)? 2 gallons
6. To shave (water running)? 10–15 gallons
7. To wash dishes by hand? 20 gallons
8. To run a dishwasher? 12 gallons

# THE ENVIROMENTAL ADDRESS BOOK GLOSSARY

## A

**Abatement:** Reducing the degree or intensity of, or eliminating, pollution.

**Abandoned Well:** A well whose use has been permanently discontinued or which is in a state of disrepair such that it cannot be used for its intended purpose.

**ABEL:** EPA's computer model for analyzing a violator's ability to pay a civil penalty.

**Absorption:** The passage of one substance into or through another (e.g., an operation in which one or more soluble components of a gas mixture are dissolved in a liquid).

**Accelerator:** In radiation science, a device that speeds up charged particles such as electrons or protons.

**Accident Site:** The location of an unexpected occurrence, failure, or loss, either at a plant or along a transportation route, resulting in a release of hazardous materials.

**Acclimatization:** The physiological and behavioral adjustments of an organism to changes in its environment.

**Acid Deposition:** A complex chemical and atmospheric phenomenon that occurs when emissions of sulfur and nitrogen compounds and other substances are transformed by chemical processes in the atmosphere, often far from the original sources, and then deposited on earth in either a wet or dry form. The wet forms, popularly called "acid rain," can fall as rain, snow, or fog. The dry forms are acidic gases or particulates.

**Acid Rain:** (see *acid deposition*)

**Activated Carbon:** A highly absorbent form of carbon used to remove odors and toxic substances from liquid or gaseous emissions. In waste treatment it is used to remove dissolved organic matter from wastewater. It is also used in motor vehicle evaporative control systems.

**Activated Sludge:** Sludge that results when primary effluent is mixed with bacteria-laden sludge and then agitated and aerated to promote biological treatment. This speeds breakdown of organic matter in raw sewage undergoing secondary waste treatment.

**Active Ingredient:** In any pesticide product, the component which kills, or otherwise controls, target pests. Pesticides are regulated primarily on the basis of active ingredients.

**Acute Exposure:** A single exposure to a toxic substance which results in severe biological harm or death. Acute exposures are usually characterized as lasting no longer than a day.

**Acute Toxicity:** The ability of a substance to cause poisonous effects resulting in severe biological harm or death soon after a single exposure or dose. Also, any severe poisonous effect resulting from a single short-term exposure to a toxic substance. (see *chronic toxicity, toxicity*)

**Adaptation:** Changes in an organism's structure or habit that help it adjust to its surroundings.

**Add-on Control Device:** An air pollution control device such as carbon absorber or incinerator, which reduces the pollution in an exhaust gas. The control device usually does not affect the process being controlled and thus is "add-on" technology, as opposed to a scheme to control pollution through making some alteration to the basic process.

**Administrative Order:** A legal document signed by the EPA directing an individual, business, or other entity to take corrective action or refrain from an activity. It describes the violations and actions to be taken, and can be enforced in court. Such orders may be issued, for example, as a result of an administrative complaint whereby the respondent is ordered to pay a penalty for violations of a statute.

**Adulterants:** Chemical impurities or substances that by law do not belong in a food or in a pesticide.

**Advanced Wastewater Treatment:** Any treatment of sewage that goes beyond the secondary or biological water treatment stage and includes the removal of nutrients such as phosphorus and nitrogen and a high percentage of suspended solids. (see *primary, secondary treatment*)

**Advisory:** A non-regulatory document that communicates risk information to persons who may have to make risk management decisions.

**Aeration:** A process which promotes biological degradation of organic water. The process may be passive (as when waste is exposed to air), or active (as when a mixing or bubbling device introduces the air).

**Aerosol:** A suspension of liquid or solid particles in a gas.

**Afterburner:** In incinerator technology, a burner located so that the combustion gases are made to pass through its flame in order to remove smoke and odors. It may be attached to or be separated from the incinerator proper.

**Agent Orange:** A toxic herbicide and defoliant which was used in the Vietnam conflict. It contains 2,4,5-trichlorophenoxyacitic acid (2,4,5-T) and 2-4 dichlorophenoxyacetic acid (2,4-D) with trace amounts of dioxin.

**Agricultural Pollution:** The liquid and solid wastes from farming, including run-off and leaching of pesticides and fertilizers; erosion and dust from plowing; animal manure and carcasses; and crop residues and debris.

**Airborne Contaminant:** Any particulate matter, gas, or combination thereof, other than water vapor or natural air. (see *air pollutant*)

**Air Curtain:** A method of containing oil spills. Air bubbling through a perforated pipe causes an upward water flow that slows the spread of oil. It can also be used to stop fish from entering polluted water.

**Air Mass:** A widespread body of air that gains certain meteorological or polluted characteristics—e.g., a heat inversion or smogginess—while set in one location. The characteristics can change as it moves away.

**Air Pollutant:** Any substance in air which could, if in high enough concentration, harm man, other animals, vegetation, or material. Pollutants may include almost any natural or artificial composition of matter capable of being airborne. They may be in the form of solid particles, liquid droplets, gases, or in combination of these forms. Generally, they fall into two main groups: 1) those emitted directly from identifiable sources and 2) those produced in the air by interaction between two or more primary pollutants, or by reaction with normal atmospheric constituents, with or without photoactivation. Exclusive of pollen, fog, and dust, which are of natural origin, about 100 contaminants have been identified and fall into the following categories: solids, sulfur compounds, volatile organic chemicals, nitrogen compounds, oxygen compounds, halogen compounds, radioactive compounds, and odors.

**Air Pollution:** The presence of contaminant or pollutant substances in the air that do not disperse properly and interfere with human health or welfare, or produce other harmful environmental effects.

**Air Pollution Episode:** A period of abnormally high concentration of air pollutants, often due to low winds and temperature inversion, that can cause illness and death. (see *episode, pollution*)

**Air Quality Control Region:** An area—designated by the federal government—in which communities share a common air pollution problem. Sometimes several states are involved.

**Air Quality Criteria:** The levels of pollution and lengths of exposure above which adverse health and welfare effects may occur.

**Air Quality Standards:** The level of pollutants prescribed by regulations that may not be exceeded during a specified time in a defined area.

**Airborne Particulates:** Total suspended particulate matter found in the atmosphere as solid particles or liquid droplets. Chemical composition of particulates varies widely, depending on location and time of year. Airborne particulates include windblown dust, emissions from industrial processes, smoke from the burning of wood and coal, and the exhaust of motor vehicles.

**Airborne Release:** Release of any chemical into the air.

**Alar:** Trade name for daminozide, a pesticide that makes apples redder, firmer, and less likely to drop off trees before growers are ready to pick them. It is also used to a lesser extent on peanuts, tart cherries, concord grapes, and other fruits.

**Algae:** Simple rootless plants that grow in sunlit waters in relative proportion to the amounts of nutrients available. They can affect water quality adversely by lowering the dissolved oxygen in the water. They are food for fish and small aquatic animals.

**Alternate Method:** Any method of sampling and analyzing for an air pollutant which is not a reference or equivalent method but which has been demonstrated in specific cases to EPA's satisfaction to produce results adequate for compliance.

**Antarctic "Ozone Hole":** Refers to the seasonal depletion of ozone in a large area over Antarctica.

**Antibodies:** Proteins produced in the body by immune system cells in response to antigens and capable of combining with antigens.

**Area Source:** Any small source of non-natural air pollution that is released over a relatively small area but which cannot be classified as a point source. Such sources may include vehicles and other small fuel combustion engines.

**Asbestos:** A mineral fiber that can pollute air or water and cause cancer or asbestosis when inhaled. EPA has banned or severely restricted its use in manufacturing and construction.

**Assimilation:** The ability of a body of water to purify itself of pollutants.

**Atmosphere (an):** A standard unit of pressure representing the pressure exerted by a 29.92-inch column of mercury at sea level at 45° latitude and equal to 1000 grams per square centimeter.

**Atmosphere (the):** The whole mass of air surrounding the earth, composed largely of oxygen and nitrogen.

# B

**Bacteria:** (singular: bacterium) Microscopic living organisms which can aid in pollution control by consuming or breaking down organic matter in sewage or by similarly acting on oil spills or other water pollutants. Bacteria in soil, water, or air can also cause human, animal, and plant health problems.

**Basal Application:** In pesticides, the application of a chemical on plant stems or tree trunks just above the soil line.

**BEN:** EPA's computer model for analyzing a violator's economic gain from not complying with the law.

**Beryllium:** An airborne metal that can be hazardous to human health when inhaled. It is discharged by machine shops, ceramic and propellant plants, and foundries.

**Bioaccumulative:** Substances that increase in concentration in living organisms (that are very slowly metabolized or excreted) as they breathe contaminated air, drink contaminated water, or eat contaminated food.

**Bio-degradable:** The ability to break down or decompose rapidly under natural conditions and processes.

**Biological Control:** In pest control, the use of animals and organisms that eat or otherwise kill or out-compete pests.

**Biotechnology:** Techniques that use living organisms or parts of organisms to produce a variety of products (from medicines to industrial enzymes) to improve plants or animals or to develop microorganisms for specific uses such as removing toxics from bodies of water, or as pesticides.

**Black Lung:** A disease of the lungs caused by habitual inhalation of coal dust.

**Botanical Pesticide:** A pesticide whose active ingredient is a plant-produced chemical such as nicotine or strychnine.

**Bottle Bill:** Proposed or enacted legislation which requires a returnable deposit on beer or soda containers and provides for retail store or other redemption centers. Such legislation is designed to discourage use of throw-away containers.

**Bottom Land Hardwoods:** Forested freshwater wetlands adjacent to rivers in the southeastern United States. They are especially valuable for wildlife breeding and nesting and habitat areas.

**Brackish Water:** A mixture of fresh and salt water.

**Burial Ground (graveyard):** A disposal site for radioactive waste materials that uses earth or water as a shield.

# C

**Carbon Dioxide ($CO_2$):** A colorless, odorless, non-poisonous gas, which results from fossil fuel combustion and is normally a part of the ambient air.

**Carbon Monoxide (CO):** A colorless, odorless, poisonous gas produced by incomplete fossil fuel combustion.

**Carcinogen:** Any substance that can cause or contribute to the production of cancer.

**Catalytic Converter:** An air pollution abatement device that removes pollutants from motor vehicle exhaust, either by oxidizing them into carbon dioxide and water or reducing them to nitrogen and oxygen.

**Catalytic Incinerator:** A control device which oxidizes volatile organic compounds (VOCs) by using a catalyst to promote the combustion process. Catalytic incinerators require lower temperatures than conventional thermal incinerators, with resultant fuel and cost savings.

**Caustic Soda:** Sodium hydroxide, a strong alkaline substance used as the cleaning agent in some detergents.

**Channelization:** Straightening and deepening streams so water will move faster, a flood-reduction or marsh-drainage tactic that can interfere with waste assimilation capacity and disturb fish and wildlife habitats.

**Chemical Treatment:** Any one of a variety of technologies that can use chemicals or a variety of chemical processes to treat waste.

**Chemosterilant:** A chemical that controls pests by preventing reproduction.

**Chilling Effect:** The lowering of the Earth's temperature because of increased particles in the air blocking the sun's rays. (see *greenhouse effect*)

**Chlorinated Hydrocarbons:** These include a class of persistent, broad-spectrum insecticides that linger in the environment and accumulate in the food chain.

Among them are DDT, aldrin, dieldrin, heptachlor, chlordane, lindane, endrin, mirex, hexachloride, and toxaphene. Other examples include TCE, used as an industrial solvent.

**Chlorination:** The application of chlorine to drinking water, sewage, or industrial waste to disinfect or to oxidize undesirable compounds.

**Chronic Toxicity:** The capacity of a substance to cause long-term poisonous human health effects. (see *acute toxicity*)

**Clarification:** Clearing action that occurs during wastewater treatment when solids settle out. This is often aided by centrifugal action and chemically induced coagulation in wastewater.

**Cleanup:** Actions taken to deal with a release or threat of release of a hazardous substance that could affect humans and/or the environment. The term "cleanup" is sometimes used interchangeably with the terms remedial action, removal action, response action, or corrective action.

**Clear Cut:** A forest management technique that involves harvesting all the trees in one area at one time. Under certain soil and slope conditions it can contribute sediment to water pollution.

**Coastal Zone:** Lands and waters adjacent to the coast that exert an influence on the uses of the sea and its ecology, or, inversely, whose uses and ecology are affected by the sea.

**Combined Sewers:** A sewer system that carries both sewage and storm-water runoff. Normally, its entire flow goes to a waste treatment plant, but during a heavy storm, the storm water volume may be so great as to cause overflows. When this happens untreated mixtures of storm water and sewage may flow into receiving waters. Storm-water run-off may also carry toxic chemicals from industrial areas or streets into the sewer system.

**Combustion:** Burning, or rapid oxidation, accompanied by release of energy in the form of heat and light. A basic cause of air pollution.

**Compliance Schedule:** A negotiated agreement between a pollution source and a government agency that specifies dates and procedures by which a source will reduce emissions and, thereby, comply with a regulation.

**Compost:** A mixture of garbage and degradable trash with soil in which certain bacteria in the soil break down the garbage and trash into organic fertilizer.

**Consent Decree:** A legal document, approved by a judge, that formalizes an agreement reached between EPA and potentially responsible parties (PRPs) through which PRPs will conduct all or part of a cleanup action at a Superfund site; cease or correct actions or processes that are polluting the environment; or otherwise comply with regulations where the PRPs' failure to comply caused EPA to initiate regulatory enforcement actions. The consent decree describes the actions PRPs will take and may be subject to a public comment period.

**Conservation:** Avoiding waste of, and renewing when possible, human and natural resources. The protection, improvement, and use of natural resources according to principles that will assure their highest economic or social benefits.

**Contact Pesticide:** A chemical that kills pests when it touches them, rather than when eaten (stomach poison). Also, soil that contains the minute skeletons of certain algae that scratch and dehydrate waxy-coated insects.

**Contaminant:** Any physical, chemical, biological, or radiological substance or matter that has an adverse affect on air, water, or soil.

**Contingency Plan:** A document setting out an organized, planned, and coordinated course of action to be followed in case of fire, explosion, or other accident that releases toxic chemicals, hazardous wastes, or radioactive materials that threaten human health or the environment.

**Criteria:** Descriptive factors taken into account by EPA in setting standards for various pollutants. These factors are used to determine limits on allowable concentration levels and limit the number of violations per year. When issued by EPA, the criteria provide guidance to the states on how to establish their standards.

# D

**DDT:** The first chlorinated hydrocarbon insecticide (chemical name: Dichloro-Diphsdyl-Trichloromethane). It has a half-life of 15 years and can collect in fatty tissues of certain animals. EPA banned registration and interstate sale of DDT for virtually all but emergency uses in the United States in 1972 because of its persistence in the environment and accumulation in the food chain.

**Decibel (dB):** A unit of sound measurement. In general, a sound doubles in loudness for every increase of ten decibels.

**Decomposition:** The breakdown of matter by bacteria and fungi. It changes the chemical makeup and physical appearance of materials.

**Defoliant:** A herbicide that removes leaves from trees and growing plants.

**Denitrification:** The anaerobic biological reduction of nitrate nitrogen to nitrogen gas.

**DES:** A synthetic estrogen, diethylstilbestrol is used as a growth stimulant in food animals. Residues in meat are thought to be carcinogenic.

**Desalinization:** Removing salt from ocean or brackish water.

**Desiccant:** A chemical agent that absorbs moisture; some desiccants are capable of drying out plants or insects, causing death.

**Designated Pollutant:** An air pollutant which is neither a criteria nor hazardous pollutant, as described in the Clean Air Act, but for which new source performance standards exist. The Clean Air Act does require states to control these pollutants, which include acid mist, total reduced sulfur (TRS), and fluorides.

**Detergent:** Synthetic washing agent that helps to remove dirt and oil. Some contain compounds which kill useful bacteria and encourage algae growth when they are in wastewater that reaches receiving waters.

**Developer:** A person, government unit, or company that proposes to build a hazardous waste treatment, storage, or disposal facility.

**Dioxin:** Any of a family of compounds known chemically as dibenzo-p-dioxins. Concern about them arises from their potential toxicity and contamination in commercial products. Tests on laboratory animals indicate that it is one of the more toxic man-made chemicals known.

**Disinfectant:** A chemical or physical process that kills pathogenic organisms in water. Chlorine is often used to disinfect sewage treatment effluents, water supplies, wells, and swimming pools.

**Disposal:** Final placement or destruction of toxic, radioactive, or other wastes; surplus or banned pesticides or other chemicals; polluted soils; and drums containing hazardous materials from removal actions or accidental releases. Disposal may be accomplished through use of approved secure landfills, surface impoundments, land farming, deep well injection, ocean dumping, or incineration.

**Dredging:** Removal of mud from the bottom of water bodies using a scooping machine. This disturbs the ecosystem and causes silting that can kill aquatic life. Dredging of contaminated muds can expose aquatic life to heavy metals and other toxics. Dredging activities may be subject to regulation under Section 404 of the Clean Water Act.

**Dump:** A site used to dispose of solid wastes without environmental controls.

# E

**Ecological Impact:** The effect that a man-made or natural activity has on living organisms and their non-living (abiotic) environment.

**Ecology:** The relationship of living things to one another and their environment, or the study of such relationships.

**Ecosphere:** The "bio-bubble" that contains life on earth, in surface waters, and in the air.

**Ecosystem:** The interacting system of a biological community and its non-living environmental surroundings.

**Effluent:** Wastewater—treated or untreated—that flows out of a treatment plant, sewer, or industrial outfall. Generally refers to wastes discharged into surface waters.

**Emergency (chemical):** A situation created by an accidental release or spill of hazardous chemicals which poses a threat to the safety of workers, residents, the environment, or property.

**Emergency Episode:** see *air pollution episode*

**Eminent Domain:** Government taking—or forced acquisition—of private land for public use, with compensation paid to the landowner.

**Emission:** Pollution discharged into the atmosphere from smokestacks, other vents, and surface areas of commercial or industrial facilities; from residential chimneys; and from motor vehicle, locomotive, or aircraft exhausts.

**Emission Standard:** The maximum amount of air polluting discharge legally allowed from a single source, mobile or stationary.

**Endangered Species:** Animals, birds, fish, plants, or other living organisms threatened with extinction by man-made or natural changes in their environment. Requirements for declaring a species endangered are contained in the Endangered Species Act.

**Endangerment Assessment:** A study conducted to determine the nature and extent of contamination at a site on the National Priorities List and the risks posed to public health or the environment. EPA or the state conducts the study when a legal action is to be taken to direct potentially responsible parties to clean up a site or pay for the cleanup. An endangerment assessment supplements a remedial investigation.

**Enforcement:** EPA, state, or local legal actions to obtain compliance with environmental laws, rules, regulations, or agreements and/or obtain penalties or criminal sanctions for violations. Enforcement procedures may vary, depending on the specific requirements of different environmental laws and related implementing regulatory requirements. Under CERCLA, for example, EPA will seek to require potentially responsible parties to clean up a Superfund site, or pay for the cleanup, whereas under the Clean Air Act the agency may invoke sanctions against cities failing to meet ambient air quality standards that could prevent certain types of construction or federal funding. In other situations, if investigations by EPA and state agencies uncover willful violations, criminal trials and penalties are sought.

**Environment:** The sum of all external conditions affecting the life, development, and survival of an organism.

**Environmental Impact Statement:** A document required of federal agencies by the National Environmental Policy Act for major projects or legislative proposals significantly affecting the environment. A tool for decision making, it describes the positive and negative effects of the undertaking and lists alternative actions.

**EPA:** The U.S. Environmental Protection Agency; established in 1970 by Presidential Executive Order, bringing together parts of various government agencies involved with the control of pollution.

**Epidemic:** Widespread outbreak of a disease, or a large number of cases of a disease in a single community or relatively small area.

**Episode (pollution):** An air pollution incident in a given area caused by a concentration of atmospheric pollution reacting with meteorological conditions that may result in a significant increase in illnesses or deaths. Although most commonly used in relation to air pollution, the term may also be used in connection with other kinds of environmental events such as a massive water pollution situation.

**Erosion:** The wearing away of land surface by wind or water. Erosion occurs naturally from weather or runoff but can be intensified by land-clearing practices related to farming, residential or industrial development, road building, or timber-cutting.

GLOSSARY

**Estuary:** Regions of interaction between rivers and nearshore ocean waters, where tidal action and river flow create a mixing of fresh and salt water. These areas may include bays, mouths of rivers, salt marshes, and lagoons. These brackish water ecosystems shelter and feed marine life, birds, and wildlife.

**Exceedance:** Violation of environmental protection standards by exceeding allowable limits or concentration levels.

**Extremely Hazardous Substance:** Any of 406 chemicals identified by EPA on the basis of toxicity and listed under SARA Title III. The list is subject to revision.

# F

**Feasibility Study:** 1. Analysis of the practicability of a proposal (e.g., a description and analysis of the potential cleanup alternatives for a site or alternatives for a site on the National Priorities List). The feasibility study usually recommends selection of a cost-effective alternative. It usually starts as soon as the remedial investigation is underway; together, they are commonly referred to as the "RI/FS." The term can apply to a variety of proposed corrective or regulatory actions. 2. In research, a small-scale investigation of a problem to ascertain whether or not a proposed research approach is likely to provide useful data.

**Fertilizer:** Materials such as nitrogen and phosphorus that provide nutrients for plants. Commercially sold fertilizers may contain other chemicals or may be in the form of processed sewage sludge.

**Filling:** Depositing dirt and mud or other materials into aquatic areas to create more dry land, usually for agricultural or commercial development purposes. Such activities often damage the ecology of the area.

**Fluorocarbons (FCs):** Any number of organic compounds analogous to hydrocarbons in which one or more hydrogen atoms are replaced by fluorine. Once used in the United States as a propellant in aerosols, they are now primarily used in coolants and some industrial processes. FCs containing chlorine are called chlorofluorocarbons (CFCs). They are believed to be modifying the ozone layer in the stratosphere, thereby allowing more harmful solar radiation to reach the Earth's surface.

**Fogging:** Applying a pesticide by rapidly heating the liquid chemical so that it forms very fine droplets that resemble smoke or fog. It may be used to destroy mosquitoes, black flies, and similar pests.

**Food Chain:** A sequence of organisms, each of which uses the next, lower member of the sequence as a food source.

**Fresh Water:** Water that generally contains less than 1,000 milligrams-per-liter of dissolved solids.

**Fuel Economy Standard:** The Corporate Average Fuel Economy Standard (CAFE) which went into effect in 1978. It was meant to enhance the national fuel conservation effort by slowing fuel consumption through a miles-per-gallon requirement for motor vehicles.

# G

**Game Fish:** Species like trout, salmon, or bass, caught for sport. Many of them show more sensitivity to environmental change than "rough" fish.

**Gasification:** Conversion of solid material such as coal into a gas for use as a fuel.

**Genetic Engineering:** A process of inserting new genetic information into existing cells in order to modify any organism for the purpose of changing one of its characteristics.

**Granular Activated-Carbon Treatment:** A filtering system often used in small water systems and individual homes to remove organics. GAC can be highly effective in removing elevated levels of radon from water.

**Gray Water:** The term given to domestic wastewater composed of washwater from sinks, bathroom sinks and tubs, and laundry tubs.

**Greenhouse Effect:** The warming of the Earth's atmosphere caused by a build-up of carbon dioxide or other traces; it is believed by many scientists that this build-up allows light from the sun's rays to heat the Earth but prevents a counterbalancing loss of heat.

**Ground Cover:** Plants grown to keep soil from eroding.

**Groundwater:** The supply of fresh water found beneath the Earth's surface (usually in aquifers) which is often used for supplying wells and springs. Because ground water is a major source of drinking water there is growing concern over areas where leaching agricultural or industrial pollutants or substances from leaking underground storage tanks are contamining ground water.

# H

**Habitat:** The place where a population (e.g., human, animal, plant, microorganism) lives and its surroundings, both living and nonliving.

**Half-life:** 1. The time required for a pollutant to lose half its affect on the environment. For example, the half-life of DDT in the environment is 15 years, of radium, 1,580 years. 2. The time required for the elimination of one half a total dose from the body.

**Hard Water:** Alkaline water containing dissolved salts that interfere with some industrial processes and prevent soap from lathering.

**Hazardous Air Pollutants:** Air pollutants which are not covered by ambient air quality standards but which, as defined in the Clean Air Act, may reasonably be expected to cause or contribute to irreversible illness or death. Such pollutants include asbestos, beryllium, mercury, benzene, coke oven emissions, radionuclides, and vinyl chloride.

**Hazardous Substance:** 1. Any material that poses a threat to human health and/or the environment. Typical hazardous substances are toxic, corrosive, ignitable, explosive, or chemically reactive. 2. Any substance named by EPA to be reported if a designated quantity of the substance is spilled in the waters of the United States or if otherwise emitted into the environment.

**Hazardous Waste:** By-products of society that can pose a substantial or potential hazard to human health or the environment when improperly managed. Possesses at least one of four characteristics (ignitability, corrosivity, reactivity, or toxicity), or appears on special EPA lists.

**Herbicide:** A chemical pesticide designed to control or destroy plants, weeds, or grasses.

**Herbivore:** An animal that feeds on plants.

**Holding Pond:** A pond or reservoir, usually made of earth, built to store polluted runoff.

**Humus:** Decomposed organic material.

**Hydrocarbons (HC):** Chemical compounds that consist entirely of carbon and hydrogen.

# I

**Ignitable:** Capable of burning or causing a fire.

**Immediately Dangerous to Life and Health (IDLH):** The maximum level to which a healthy individual can be exposed to a chemical for 30 minutes and escape without suffering irreversible health effects or impairing symptoms. Used as a "level of concern."

**Incineration:** 1. Burning of certain types of solid, liquid, or gaseous materials. 2. A treatment technology involving destruction of waste by controlled burning at high temperatures (e.g., burning sludge to remove the water and reduce the remaining residues to a safe, non-burnable ash which can be disposed of safely on land, in some waters, or in underground locations).

**Indirect Discharge:** Introduction of pollutants from a non-domestic source into a publicly owned waste treatment system. Indirect dischargers can be commercial or industrial facilities whose wastes go into the local sewers.

**Indoor Air:** The breathing air inside a habitable structure or conveyance.

**Indoor Climate:** Temperature, humidity, lighting, and noise levels in a habitable structure or conveyance. Indoor climate can affect indoor air pollution.

**Inert Ingredient:** Pesticide components such as solvents, carriers, and surfactants that are not active against target pests. Not all inert ingredients are innocuous.

**Infiltration:** 1. The penetration of water through the ground surface into subsurface soil or the penetration of water from the soil into sewer or other pipes through defective joints, connections, or manhole walls. 2. A land application technique where large volumes of wastewater are applied to land, allowed to penetrate the surface and percolate through the underlying soil.

**Information File:** In the Superfund program, a file that contains accurate, up-to-date documents on a Superfund site. The file is usually located in a public building such as a school, library, or city hall that is convenient for local residents.

**Inorganic Chemicals:** Chemical substances of mineral origin, not of basically carbon structure.

**Insecticide:** A pesticide compound specifically used to kill or control the growth of insects.

**Inspection and Maintenance (I/M):** 1. Activities to assure proper emissions-related operation of mobile sources of air pollutants, particularly automobile emissions controls. 2. Also applies to wastewater treatment plants and other anti-pollution facilities and processes.

**Interim (permit) Status:** Period during which treatment, storage, and disposal facilities coming under RCRA in 1980 are temporarily permitted to operate while awaiting denial or issuance of a permanent permit. Permits issued under these circumstances are usually called "Part A" or "Part B" permits.

**Inversion:** An atmospheric condition caused by a layer of warm air preventing the rise of cooling air trapped beneath it. This prevents the rise of pollutants that might otherwise be dispersed and can cause an air pollution episode.

**Irradiated Food:** Food that has been subject to brief radioactivity, usually by gamma rays, to kill insects, bacteria, and mold, and preserve it without refrigeration or freezing.

**Irradiation:** Exposure to radiation of wavelengths shorter than those of visible light (gamma, x-ray, or ultraviolet), for medical purposes, the destruction of bacteria in milk or other foodstuffs, or for inducing polymerization of monomers or vulcanization of rubber.

**Irrigation:** Technique for applying water or wastewater to land areas to supply the water and nutrient needs of plants.

# L

**Lagoon:** 1. A shallow pond where sunlight, bacterial action, and oxygen work to purify wastewater; also used for storage of wastewaters or spent nuclear fuel rods. 2. Shallow body of water, often separated from the sea by coral reefs or sandbars.

**Land Application:** Discharge of wastewater onto the ground for treatment or re-use.

**Land Farming (of waste):** A disposal process in which hazardous waste deposited on or in the soil is naturally degraded by microbes.

**Landfills:** 1. Sanitary landfills are land disposal sites for non-hazardous solid wastes at which the waste is spread in layers, compacted to the smallest practical volume, and cover material applied at the end of each operating day. 2. Secure chemical landfills are disposal sites for hazardous waste. They are selected and designed to minimize the chance of release of hazardous substances into the environment.

**Leachate:** A liquid that results from water collecting contaminants as it trickles through wastes, agricultural pesticides, or fertilizers. Leaching may occur in farming areas, feedlots, and landfills, and may result in hazardous substances entering surface water, ground water, or soil.

**Leaded Gasoline:** Gasoline to which lead has been added to raise the octane level.

**Level of Concern (LOC):** The concentration in air of an extremely hazardous substance above which there may be serious immediate health effects to anyone exposed to it for short periods of time.

**Liner:** 1. A relatively impermeable barrier designed to prevent leachate from leaking from a landfill. Liner materials include plastic and dense clay. 2. An insert or sleeve for sewer pipes to prevent leakage or infiltration.

**Liquefaction:** Changing a solid into a liquid.

**List:** Shorthand term for EPA list of violating facilities or list of firms debarred from obtaining government contracts because they violated certain sections of the Clean Air or Clean Water Acts. The list is maintained by the Office of Enforcement and Compliance Monitoring.

**Listed Waste:** Wastes listed as hazardous under RCRA but which have not been subjected to the Toxic Characteristics Listing Process because the dangers they present are considered self-evident.

**Local Emergency Planning Committee (LEPC):** A committee appointed by the state emergency response commission, as required by SARA Title III, to formulate a comprehensive emergency plan for its jurisdiction.

**Lowest Achievable Emission Rate:** Under the Clean Air Act, this is the rate of emissions which reflects (a) the most stringent emission limitation which is contained in the implementation plan of any state for such source unless the owner or operator of the proposed source demonstrates such limitations are not achievable; or (b) the most stringent emissions limitation achieved in practice, whichever is more stringent. Application of this term does not permit a proposed new or modified source to emit pollutants in excess of existing new source standards.

# M

**Marsh:** A type of wetland that does not accumulate appreciable peat deposits and is dominated by herbaceous vegetation. Marshes may either be fresh or saltwater and tidal or non-tidal.

**Maximum Contaminant Level:** The maximum permissible level of a contaminant in water delivered to any user of a public water system. MCLs are enforceable standards.

**Methane:** A colorless, non-poisonous, flammable gas created by anaerobic decomposition of organic compounds.

**Method 18:** An EPA reference test method which uses gas chromatographic techniques to measure the concentration of individual volatile organic compounds in a gas stream.

**Microbes:** Microscopic organisms such as algae, animals, viruses, bacteria, fungus, and protozoa, some of which cause diseases.

**Mobile Source:** A moving producer of air pollution, mainly forms of transportation such as cars, trucks, motorcycles, airplanes.

**Monitoring:** Periodic or continuous surveillance or testing to determine the level of compliance with statutory requirements and/or pollutant levels in various media or in humans, animals, and other living things.

**Muck Soils:** Earth made from decaying plant materials.

**Mulch:** A layer of material (wood chips, straw, leaves, etc.) placed around plants to hold moisture, prevent weed growth, protect plants, and enrich soil.

**Multiple Use:** Use of land for more than one purpose (i.e., grazing of livestock, wildlife production, recreation, watershed, and timber production). Could also apply to use of bodies of water for recreational purposes, fishing, and water supply.

# N

**National Ambient Air Quality Standards (NAAQS):** Air quality standards established by EPA that apply to outside air throughout the country.

**National Emissions Standards for Hazardous Air Pollutants (NESHAPS):** Emissions standards set by EPA for an air pollutant not covered by NAAQS that may cause an increase in deaths or in serious, irreversible, or incapacitating illness. Primary standards are designed to protect human health, secondary standards to protect public welfare.

**National Oil and Hazardous Substances Contingency Plan (NOHSCP/NCP):** The federal regulation that guides determination of the sites to be corrected under the Superfund program and the program to prevent or control spills into surface waters or other portions of the environment.

**National Pollutant Discharge Elimination System (NPDES):** A provision of the Clean Water Act which prohibits discharge of pollutants into waters of the United States unless a special permit is issued by EPA, a state, or (where delegated) a tribal government on an Indian reservation.

**National Priorities List (NPL):** EPA's list of the most serious uncontrolled or abandoned hazardous waste sites identified for possible long-term remedial action under Superfund. A site must be on the NPL to receive money from the Trust Fund for remedial action. The list is based primarily on the score a site receives from the Hazard Ranking System. EPA is required to update the NPL at least once a year.

**National Response Center:** The federal operations center that receives notifications of all releases of oil and hazardous substances into the environment. The Center, open 24 hours a day, is operated by the U.S. Coast Guard, which evaluates all reports and notifies the appropriate agency.

**National Response Team (NRT):** Representatives of 13 federal agencies that, as a team, coordinate federal responses to nationally significant incidents of pollution and provide advice and technical assistance to the responding agency (or agencies) before and during a response action.

**Navigable Waters:** Traditionally, waters sufficiently deep and wide for navigation by all, or specified sizes of vessels; such waters in the United States come under federal jurisdiction and are included in certain provisions of the Clean Water Act.

**New Source Performance Standards (NSPS):** Uniform national EPA air emission and water effluent standards which limit the amount of pollution allowed from new sources or from existing sources that have been modified.

**Nitrate:** A compound containing nitrogen which can exist in the atmosphere or as a dissolved gas in water and which can have harmful effects on humans and animals. Nitrates in water can cause severe illness in infants and cows.

**Nitrogen Dioxide ($NO_2$):** The result of nitric oxide combining with oxygen in the atmosphere. A major component of photochemical smog.

**Nuclear Power Plant:** A facility that converts atomic energy into usable power; heat produced by a reactor makes steam to drive turbines which produce electricity.

**Nuclear Winter:** Prediction by some scientists that smoke and debris rising from massive fires resulting from a nuclear war could enter the atmosphere and block out sunlight for weeks or months. The scientists making this prediction project a cooling of the earth's surface and changes in climate which could, for example, negatively affect world agricultural and weather patterns.

# O

**Off-Site Facility:** A hazardous waste treatment, storage, or disposal area that is located at a place away from the generating site.

**Oil Spill:** An accidental or intentional discharge of oil which reaches bodies of water; can be controlled by chemical dispersion, combustion, mechanical containment, and/or absorption.

**On-Site Facility:** A hazardous waste treatment, storage, or disposal area that is located on the generating site.

**Open Burning:** Uncontrolled fires in an open dump.

**Open Dump:** An uncovered site used for disposal of waste without environmental controls.

**Organism:** Any living thing.

**Oxidant:** A substance containing oxygen that reacts chemically in air to produce a new substance. The primary ingredient of photochemical smog.

**Oxidation:** 1. The addition of oxygen, which breaks down organic waste or chemicals such as cyanides, phenols, and organic sulfur compounds in sewage by bacterial and chemical means. 2. Oxygen combining with other elements. 3. The process in chemistry whereby electrons are removed from the molecule.

**Ozone ($O_3$):** Found in two layers of the atmosphere, the stratosphere and the troposphere. In the stratosphere (the atmospheric layer beginning 7 to 10 miles above the earth's surface), ozone is a form of oxygen found naturally, which provides a protective layer shielding the earth from ultraviolet radiation's harmful health effects on humans and the environment. In the troposphere (the layer extending up 7 to 10 miles from the earth's surface), ozone is a chemical oxidant and major component of photochemical smog. Ozone can seriously affect the human respiratory system and is one of the most prevalent and widespread of all the criteria pollutants for which the Clean Air Act required EPA to set standards. Ozone in the troposphere is produced through: complex chemical reactions of nitrogen oxides, which are among the primary pollutants emitted by combustion sources; hydrocarbons, released into the atmosphere through the combustion, handling, and processing of petroleum products; and sunlight.

**Ozone Depletion:** Destruction of the stratospheric ozone layer, which shields the earth from ultraviolet radiation harmful to biological life. This destruction of ozone is caused by the breakdown of certain chlorine- and/or bromine-containing compounds (chlorofluorocarbons or halons) that break down when they reach the stratosphere and catalytically destroy ozone molecules.

# P

**Paraquat:** A standard herbicide used to kill various types of crops, including marijuana.

**Pathogens:** Microorganisms that can cause disease in other organisms or in humans, animals, and plants. They may be bacteria, viruses, or parasites and are found in sewage, in run-off from animal farms or rural areas populated with domestic and/or wild animals, and in water used for swimming. Fish and shellfish contaminated by pathogens, or the contaminated by pathogens, or the contaminated water itself, can cause serious illnesses.

**Pest:** An insect, rodent, nematode, fungus, weed, or other form of terrestrial or aquatic plant or animal life or virus, bacterial or microorganism, that is injurious to health or the environment.

**Pesticide:** Substance or mixture of substances intended for preventing, destroying, repelling, or mitigating any pest. Also, any substance or mixture of substances intended for use as a plant regulator, defoliant, or desiccant. Pesticides can accumulate in the food chain and/or contaminate the environment if misused.

**Photochemical Smog:** Air pollution caused by chemical reactions of various pollutants emitted from different sources.

**Photosynthesis:** The manufacture by plants of carbohydrates and oxygen from carbon dioxide and water in the presence of chlorophyll, using sunlight as an energy source.

**Plankton:** Tiny plants and animals that live in water.

**Plastics:** Non-metallic compounds that result from a chemical reaction and are molded or formed into rigid or pliable construction materials or fabrics.

**Plume:** 1. A visible or measurable discharge of a contaminant from a given point of origin; can be visible or thermal in water, or visible in the air as, for example, a plume of smoke. 2. The area of measurable and potentially harmful radiation leaking from a damaged reactor. 3. The distance from a toxic release considered dangerous for those exposed to the leaking fumes.

**Point Source:** A stationary location or fixed facility from which pollutants are discharged or emitted. Also, any single identifiable source of pollution (e.g., a pipe, ditch, ship, ore pit, factory smokestack).

**Pollen:** 1. A fine dust produced by plants. 2. The fertilizing element of flowering plants. 3. A natural or background air pollutant.

**Pollutant:** Generally, any substance introduced into the environment that adversely affects the usefulness of a resource.

**Pollution:** Generally, the presence of matter or energy whose nature, location, or quantity produces undesired environmental effects. Under the Clean Water Act, for example, the term is defined as the man-made or man-induced alteration of the physical, biological, and radiological integrity of water.

**Population:** A group of interbreeding organisms of the same kind occupying a particular space. Generically, the number of humans or other living creatures in a designated area.

**Potable Water:** Water that is safe for drinking and cooking.

**Potentially Responsible Party (PRP):** Any individual or company—including owners, operators, transporters, or generators—potentially responsible for, or contributing to, the contamination problems at a Superfund site. Whenever possible, EPA requires PRPs, through administrative and legal actions, to clean up hazardous waste sites PRPs have contaminated.

**Pretreatment:** Processes used to reduce, eliminate, or alter the nature of wastewater pollutants from non-domestic sources before they are discharged into publicly owned treatment works.

**Prevention:** Measures taken to minimize the release of wastes to the environment.

**Primary Drinking Water Regulation:** Applies to public water systems and specifies a contaminant level, which, in the judgment of the EPA Administrator, will have no adverse effect on human health.

**Primary Waste Treatment:** First steps in wastewater treatment; screens and sedimentation tanks are used to remove most materials that float or will settle. Primary treatment results in the removal of about 30% of carbonaceous biochemical oxygen demand from domestic sewage.

# Q

**Quality Assurance/Quality Control:** A system of procedures, checks, audits, and corrective actions to ensure that all EPA research design and performance, environmental monitoring and sampling, and other technical and reporting activities are of the highest achievable quality.

# R

**Radiation:** Any form of energy propagated as rays, waves, or streams of energetic particles. The term is frequently used in relation to the emission of rays from the nucleus of an atom.

**Radiation Standards:** Regulations that set maximum exposure limits for protection of the public from radioactive materials.

**Radioactive Substances:** Substances that emit radiation.

**Radon:** A colorless, naturally occurring, radioactive, inert gaseous element formed by radioactive decay of radium atoms in soil or rocks.

**Raw Sewage:** Untreated wastewater.

**Reasonably Available Control Technology (RACT):** The lowest emissions limit that a particular source is capable of meeting by the application of control technology that is both reasonably available, as well as technologically and economically feasible. RACT is usually applied to existing sources in non-attainment areas and in most cases is less stringent than new source performance standards.

**Recommended Maximum Contaminant Level (RMCL):** The maximum level of a contaminant in drinking water at which no known or anticipated adverse effect on human health would occur, and which includes an adequate margin of safety. Recommended levels are non-enforceable health goals.

**Recycle/Re-use:** The process of minimizing the generation of waste by recovering usable products that might otherwise become waste. Examples are the recycling of aluminum cans, wastepaper, and bottles.

**Refuse Reclamation:** Conversion of solid waste into useful products (e.g., composting organic wastes to make soil conditioners or separating aluminum and other metals for melting and recycling).

**Reportable Quantity (RQ):** The quantity of a hazardous substance that triggers reports under CERCLA. If a substance is released in amounts exceeding its RQ, the release must be reported to the National Response Center, the State Emergency Response Commission, and community emergency coordinators for areas likely to be affected.

**Reservoir:** Any natural or artificial holding area used to store, regulate, or control water.

**Residual:** Amount of a pollutant remaining in the environment after a natural or technological process has taken place (e.g., the sludge remaining after initial wastewater treatment, or particulates remaining in air after the air passes through a scrubbing or other pollutant-removal process).

**Resistance:** For plants and animals, the ability to withstand poor environmental conditions and/or attacks by chemicals or disease. The ability may be inborn or developed.

**Resource:** A person, thing, or action needed for living or to improve the quality of life.

**Resource Recovery:** The process of obtaining matter or energy from materials formerly discarded.

**Restoration:** Measures taken to return a site to pre-violation conditions.

**Ringlemann Chart:** A series of shaded illustrations used to measure the opacity of air pollution emissions. The chart ranges from light gray through black and is used to set and enforce emissions standards.

**Risk Assessment:** The qualitative and quantitative evaluation performed in an effort to define the risk posed to human health and/or the environment by the presence or potential presence and/or use of specific pollutants.

**Risk Management:** The process of evaluating alternative regulatory and non-regulatory responses to risk and selecting among them. The selection process necessarily requires the consideration of legal, economic, and social factors.

**Rodenticide:** A chemical or agent used to destroy rats or other rodent pests, or to prevent them from damaging food, crops, etc.

**Rough Fish:** Those fish, not prized for eating, such as gar and suckers. Most are more tolerant of changing environmental conditions than game species.

**Rubbish:** Solid waste, excluding food waste and ashes, from homes, institutions, and workplaces.

**Run-Off:** That part of precipitation, snow-melt, or irrigation water that runs off the land into streams or other surface-water. It can carry pollutants from the air and land into receiving waters.

## S

**Salinity:** The degree of salt in water.

**Salvage:** The utilization of waste materials.

**Sanitation:** Control of physical factors in the human environment that could harm development, health, or survival.

**Secondary Drinking Water Regulations:** Unenforceable regulations which apply to public water systems and which specify the maximum contamination levels which, in the judgement of EPA, are required to protect the public welfare. These regulations apply to any contaminants that may adversely affect the odor or appearance of such water and consequently may cause people served by the system to discontinue its use.

**Secondary Wastewater Treatment:** The second step in most publicly owned waste treatment systems in which bacteria consume the organic parts of the waste. It is accomplished by bringing together waste, bacteria, and oxygen in trickling filters or in the activated sludge process. This treatment removes floating and settleable solids and about 90% of the oxygen-demanding substances and suspended solids. Disinfection is the final stage of secondary treatment.

**Sediments:** Soil, sand, and minerals washed from land into water usually after rain. They pile up in reservoirs, rivers, and harbors, destroying fish-nesting areas and holes of water animals and clouding the water so that needed sunlight might not reach aquatic plants. Careless farming, mining, and building activities will expose sediment materials, allowing them to be washed off the land after rainfalls.

**Selective Pesticide:** A chemical designed to affect only certain types of pests, leaving other plants and animals unharmed.

**Septic Tank:** An underground storage tank for wastes from homes having no sewer line to a treatment plant. The waste goes directly from the home to the tank, where the organic waste is decomposed by bacteria and the sludge settles to the bottom. The effluent flows out of the tank into the ground through drains; the sludge is pumped out periodically.

**Sewage:** The waste and wastewater produced by residential and commercial establishments and discharged into sewers.

**Sewage Sludge:** Sludge produced at a Publicly Owned Treatment Works, the disposal of which is regulated under the Clean Water Act.

**Sewer:** A channel or conduit that carries wastewater and storm water run-off from the source to a treatment plant or receiving stream. Sanitary sewers carry household, industrial, and commercial waste. Storm sewers carry run-off from rain or snow. Combined sewers are used for both purposes.

**Significant Violations:** Violations of government pollution standards that are so important in size or duration that their correction becomes a high priority.

**Silt:** Fine particles of sand or rock that can be picked up by the air or water and deposited as sediment.

**Sludge:** A semi-solid residue from any of a number of air or water treatment processes. Sludge can be hazardous waste.

**Smog:** Air pollution associated with oxidants.

**Smoke:** Particles suspended in air after incomplete combustion of materials.

**Soft Detergents:** Cleaning agents that break down in nature.

**Soft Water:** Any water that is not "hard" (i.e., does not contain a significant amount of dissolved minerals such as salts containing calcium or magnesium).

**Solid Waste:** Non-liquid, non-soluble materials ranging from municipal garbage to industrial wastes that contain complex, and sometimes hazardous, substances. Solid wastes also include sewage sludge, agricultural refuse, demolition wastes, and mining residues. Technically, solid waste also refers to liquids and gases in containers.

**Solid Waste Disposal:** The final placement of refuse that is not salvaged or recycled.

**Solid Waste Management:** Supervised handling of waste materials from their source through recovery processes to disposal.

**Soot:** Carbon dust formed by incomplete combustion.

**Sprawl:** Unplanned development of open land.

**Stabilization:** Conversion of the active organic matter in sludge into inert, harmless material.

**Stable Air:** A mass of air that is not moving normally, so that it holds rather than disperses pollutants.

**State Emergency Response Commission (SERC):** Commission appointed by each state governor according to the requirements of SARA Title III. The SERCs designate emergency planning districts, appoint local emergency planning committees, and supervise and coordinate their activities.

**Stationary Source:** A fixed, non-moving producer of pollution, mainly power plants and other facilities using industrial combustion processes.

**Sterilization:** 1. In pest control, the use of radiation and chemicals to damage body cells needed for reproduction. 2. The destruction of all living organisms in water or on the surface of various materials. In contrast, disinfection is the destruction of most living organisms in water or on surfaces.

**Storm Sewer:** A system of pipes (separate from sanitary sewers) that carry only water run-off from building and land surfaces.

**Strip-Cropping:** Growing crops in a systematic arrangement of strips or bands which serve as barriers to wind and water erosion.

**Strip-Mining:** A process that uses machines to scrape soil or rock away from mineral deposits just under the earth's surface.

**Sulfur Dioxide (SO$_2$):** A heavy, pungent, colorless, gaseous air pollutant formed primarily by industrial fossil fuel combustion processes.

**Surface Water:** All water naturally open to the atmosphere (rivers, lakes, reservoirs, streams, impoundments, seas, estuaries, etc.); also refers to springs, wells, or other collectors which are directly influenced by surface water.

**Suspension:** The act of suspending the use of a pesticide when EPA deems it necessary to do so in order to prevent an imminent hazard resulting from continued use of the pesticide. An emergency suspension takes effect immediately; under an ordinary suspension a registrant can request a hearing before the suspension goes into effect. Such a hearing process might take six months.

**Swamp:** A type of wetland that is dominated by woody vegetation and does not accumulate appreciable peat deposits. Swamps may be fresh or salt water and tidal or non-tidal.

**Synergism:** The cooperative interaction of two or more chemicals or other phenomena producing a greater total effect than the sum of their individual effects.

# T

**Tertiary Treatment:** Advanced cleaning of wastewater that goes beyond the secondary or biological stage. It removes nutrients such as phosphorus and nitrogen and most BOD and suspended solids.

**Thermal Pollution:** Discharge of heated water from industrial processes that can affect the life processes of aquatic organisms.

**Tolerances:** The permissible residue levels for pesticides in raw agricultural produce and processed foods. Whenever a pesticide is registered for use on a food or feed crop, a tolerance (or exemption from the tolerance requirement) must be established. EPA establishes the tolerance levels, which are enforced by the Food and Drug Administration and the Department of Agriculture.

**Toxic:** Harmful to living organisms.

**Toxic Cloud:** Airborne mass of gases, vapors, fumes, or aerosols containing toxic materials.

**Toxic Pollutants:** Materials contaminating the environment that cause death, disease, and/or birth defects in organisms that ingest or absorb them. The quantities and length of exposure necessary to cause these effects can vary widely.

**Toxic Substance:** A chemical or mixture that may present an unreasonable risk of injury to health or the environment.

**Toxicant:** A poisonous agent that kills or injuries animal or plant life.

**Toxicity:** The degree of danger posed by a substance to animal or plant life.

**Toxicology:** The science and study of poisons control.

**Trash-to-Energy Plan:** A plan for putting waste back to work by burning trash to produce energy.

**Treatment, Storage, and Disposal Facility:** Site where a hazardous substance is treated, stored, or disposed. TSD facilities are regulated by EPA and states under RCRA.

**Tundra:** A type of ecosystem dominated by lichens, mosses, grasses, and woody plants. Tundra is found at high latitudes (arctic tundra) and high altitudes (alpine tundra). Arctic tundra is underlain by permafrost and is usually very wet.

**Turbidity:** 1. Haziness in air caused by the presence of particles and pollutants. 2. A similar cloudy condition in water due to suspended silt or organic matter.

# U

**Ultraviolet Rays:** Radiation from the sun that can be useful or potentially harmful. UV rays from one part of the spectrum enhance plant life and are useful in some medical and dental procedures; UV rays from other parts of the spectrum to which humans are exposed (e.g., while getting a sun tan) can cause skin cancer or other tissue damage. The ozone layer in the atmosphere provides a protective shield that limits the amount of ultraviolet rays that reach the Earth's surface.

**Underground Sources of Drinking Water:** As defined in the UIC program, this term refers to aquifers that are currently being used as a source of drinking water, and those that are capable of supplying a public water system. They have a total dissolved solids content of 10,000 milligrams per liter or less, and are not "exempted aquifers."

# V

**Vaccine:** Dead, partial, or modified antigen used to induce immunity to certain infectious diseases.

**Vapor:** The gaseous phase of substances that are liquid or solid at atmospheric temperature and pressure (e.g., steam).

**Vaporization:** The change of a substance from a liquid to a gas.

**Variance:** Government permission for a delay or exception in the application of a given law, ordinance, or regulation.

**Virus:** The smallest form of microorganisms capable of causing disease.

**Vulnerable Zone:** An area over which the airborne concentration of a chemical involved in an accidental release could reach the level of concern.

# W

**Waste:** 1. Unwanted materials left over from a manufacturing process. 2. Refuse from places of human or animal habitation.

**Waste Treatment Plant:** A facility containing a series of tanks, screens, filters, and other processes by which pollutants are removed from water.

**Water Pollution:** The presence in water of enough harmful or objectionable material to damage the water's quality.

**Water Quality Standards:** State-adopted and EPA-approved ambient standards for water bodies. The standards cover the use of the water body and the water quality criteria which must be met to protect the designated use or uses.

**Water Table:** The level of ground water.

**Watershed:** The land area that drains onto a stream.

**Wetlands:** An area that is regularly saturated by surface or ground water and subsequently is characterized by a prevalence of vegetation that is adapted for life in saturated soil conditions. Examples include swamps, bogs, fens, marshes, and estuaries.

**Wildlife Refuge:** An area designated for the protection of wild animals, within which hunting and fishing are either prohibited or strictly controlled.

**Wood-Burning Stove Pollution:** Air pollution caused by emissions of particulate matter, carbon monoxide, total suspended particulates, and polycyclic organic matter from wood-burning stoves.

# X,Y,Z

**Xenobiotic:** Term for non-naturally occurring man-made substances found in the environment (i.e., synthetic material solvents, plastics).

**Zooplankton:** Tiny aquatic animals eaten by fish.

Also by Michael Levine:

# THE ADDRESS BOOK
*How to Reach Anyone Who* Is *Anyone*

This definitive address book lists current names and addresses for over 3,500 VIPs and celebrities in every field imaginable, now in its fifth edition—a Perigee classic.

This book is available at your local bookstore or wherever books are sold. Also, ordering is easy and convenient. Just call 1-800-631-8571 or send your order to:

The Putnam Publishing Group
390 Murray Hill Parkway, Dept. B
East Rutherford, NJ 07073

---

|  |  | U.S. | CANADA |
|---|---|---|---|
| _____ THE ADDRESS BOOK | 399-51621 | $9.95 | $12.95 |

|  |  |
|---|---|
| Subtotal | $ _____ |
| *Postage & handling | $ _____ |
| Sales Tax | $ _____ |
| (CA, NJ, NY, PA) |  |
| Total Amount Due | $ _____ |
| Payable in U.S. Funds |  |
| (No cash orders accepted) |  |

*Postage & handling: $2.00 for 1 book, 50¢ for each additional book up to a maximum of $4.50

- - - - - - - - - - - - - - - - - - - - - - - - - - - - - - - - - - - - - - - - - - - - - - - -

Enclosed is my   ☐ check   ☐ money order
Please charge my   ☐ Visa   ☐ MasterCard   ☐ American Express
Card # _____ Expiration date _____
Minimum order for credit cards is $10.00.
Signature as on charge card _____
Name _____
Address _____
City _____ State _____ Zip _____

Please allow six weeks for delivery. Prices subject to change without notice.